Practical Handbook of

Disturbed Land Revegetation

Practical Handbook of
Disturbed Land Revegetation

Frank F. Munshower
Reclamation Research Unit
Montana State University
Bozeman, Montana

CRC Press
Taylor & Francis Group
Boca Raton London New York

CRC Press is an imprint of the
Taylor & Francis Group, an **informa** business

First published 1994 by CRC Press
Taylor & Francis Group
6000 Broken Sound Parkway NW, Suite 300
Boca Raton, FL 33487-2742

Reissued 2018 by CRC Press

Library of Congress Cataloging-in-Publication Data

Munshower, Frank F.
Practical handbook of disturbed land revegetation / Frank F. Munshower.
p. cm.
Includes bibliographical references and index.
ISBN 1-56670-026-4
1. Revegetation—Handbooks, manuals, etc. 2. Reclamation of land—Handbooks, manuals, etc. I. Title. II. Title: Disturbed land revegetation.
S627.R47M86 1993
631.6'4—dc20 93-21625

A Library of Congress record exists under LC control number: 93021625

ISBN 13: 978-1-315-89682-3 (hbk)
ISBN 13: 978-1-351-07592-3 (ebk)

Visit the Taylor & Francis Web site at http://www.taylorandfrancis.com and the
CRC Press Web site at http://www.crcpress.com

ACKNOWLEDGMENTS

This text is a compilation of the writings and thoughts of a number of rehabilitation scientists and practitioners. It was originally written as a set of class notes, then as a series of class handouts, and finally as a text for general distribution. To acknowledge all of the reclamation specialists, mine company personnel, scientists, and ranchers who have made direct contributions to this text would be impossible. The references cited acknowledge some contributors, but the list is far from complete.

This book is also the result of many years of mined land rehabilitation research and demonstration by the author. This activity was funded by a number of private corporations, Montana state agencies, and federal organizations. Chief among these were the Montana Power Company, the Western Energy Company, Atlantic Richfield, the U.S. Bureau of Mines, the Office of Surface Mining, the Montana Department of State Lands, and the Montana Department of Natural Resources and Conservation. The Western Energy Company must be mentioned because of its continued dedication to outstanding reclamation and a long standing program of rehabilitation research at the Rosebud Mine in southeastern Montana. While other companies have produced more glamorous demonstrations of rehabilitation on small parcels of land, the Rosebud Mine has reproduced native rangeland complete with wildlife, windmills pumping stock water from spoil aquifers, hay meadows, fields of grain, and natural-appearing impoundments on thousands of acres of western minesoils. At the same time, they have continued their program of mined land rehabilitation research into means of creating better and more cost-effective reclamation and of achieving complete restoration of their minesoils.

Note must be made of the editorial efforts of Gerald Schuman of the Agricultural Research Service in Sheridan, Wyoming and Joyce Doherty of Chevy Chase, Maryland. Their reviews of this document were thorough and undoubtedly time-consuming, but their efforts have improved the text grammatically and factually.

THE AUTHOR

Frank F. Munshower, Ph.D., is Director of the Reclamation Research Unit, Montana State University, Bozeman, Montana.

After graduation from Purdue University, Dr. Munshower taught high school biology in Indianpolis, Indiana for almost ten years. He then returned to graduate school at the University of Montana where he received his Ph.D. in Botany in 1972. His research career with the Reclamation Research Unit at Montana State University coincided with the beginning of large scale strip mining of coal on the Great Plains.

Dr. Munshower has been occupied with research in various phases of land rehabilitation since that time. After his appointment as head of the Reclamation Research Unit in 1980, he was instrumental in the establishment of a master's degree program in Land Rehabilitation. An undergraduate degree in abused land rehabilitation was added in 1993. In 1992 Dr. Munshower was designated as the Outstanding Researcher of the Year by the American Society of Surface Mining and Reclamation.

His research has included a number of types of degraded lands. These have included highway shoulders, coal minesoils, metal mine wastes, ash disposal ponds, tailings, and inorganic hazardous waste sites. He has co-chaired the biennial Billings, Montana Reclamation Symposium with the Western Technical Center of the Office of Surface Mining for over ten years.

TABLE OF CONTENTS

Introduction 1

References 4

Chapter 1. Soil Parameters That Directly Affect Plant Establishment and Development on Disturbed Lands 5

Hydrogen Ion Concentration 5

Electrical Conductivity 6

Sodium in Soils 7

Organic Matter 8

Texture and Structure 10

Bulk Density and Porosity 10

Cation Exchange Capacity 12

Other Soil Tests 13

References 14

Chapter 2. Plant and Soil Elemental Chemistry 15

Macronutrients 17
- Nitrogen 18
- Phosphorus 20
- Potassium 21
- Calcium 21
- Magnesium 22
- Sulfur 23

Trace Elements 24
- Aluminum 26
- Arsenic 27

Boron 29
Cadmium 30
Chlorine 31
Copper 32
Iron 33
Lead 34
Manganese 35
Mercury 36
Molybdenum 37
Nickel 38
Selenium 39
Sodium 43
Zinc 43

References 44

Chapter 3. Site Preparation 57

Recontouring 58
Slope Reconstruction 60
Integrity of Drainage Basins 64

Topsoiling 66

Surface Manipulation 69
Standard Farming Techniques 70
Nontypical Seedbed Preparation 72

Amendments 76
Fertilizers 79
Nitrogen Fertilization 80
Phosphorus Fertilization 82
Potassium Fertilization 83
Fertilizer Mechanics 84
Mulches and Organic Amendments 85
Paper 89
Wood Residues 89
Straw 90
Native Hay 91
Stubble Mulch 92
Erosion Control Blankets 92
Manure 94
Green Manure 94
Compost 95
Sewage Sludge 96

Lime 97
Sodic Soil Amendments 99

Irrigation 100

References 101

Chapter 4. Seeding and Planting 109

Seedbed Preparation 111

Seed Mixes 112
Plant Species Selection 113
Seed Rates 117

Seeding Techniques 118
Drill Seeding 119
Broadcast Seeding 122

Seeding Legumes 124

Seeding Temporary Stabilizing Species 125

Season of Seeding 127

Planting 128
Whole Plants 128
Plant Parts 133
Special Planting Techniques 133
Planting Failure 136

References 138

Chapter 5. Special Reclamation Practices to Enhance Vegetation Diversity and Usefulness 141

Community Diversity 144

Landscape Diversity 148
Variable Depth Topsoiling 150
Bluff Extensions 151
Impoundments 152
Inplace Spoil Ridges 156

Rock Ledges 157

References 159

Chapter 6. Applied Revegetation Techniques for Common Disturbances 161

Western Coal Mining 161

Metal Mine Wastes 167
Open Pit 169
Waste Rock 170
Access Roads 176
Tailings 178

Western Phosphate Mining 188

Western Bentonite Mining 191

Sand and Gravel Pits 195

Highway Shoulders 198

Ash Disposal Ponds 201

Drill Pads 204

Abandoned Mine Lands (AML) 204

References 207

Appendix. Plant Species Commonly Used in Revegetation Programs 213

References 226

Index 229

LIST OF TABLES

Table 1. Soil reaction ranges........ 6
Table 2. Soil salinity guide........ 6
Table 3. Suggested ESP and SAR ranges........ 8
Table 4. Soil organic matter levels........ 8
Table 5. Diameter range of the primary soil particles........ 10
Table 6. The range of soil bulk density and porosity........ 11
Table 7. Soil permeability rates........ 12
Table 8. General ranges of soil CEC........ 13
Table 9. Plant nutrients........ 16
Table 10. Trace element ranges common in soils........ 25
Table 11. Livestock response to Cu and Mo in their diet........ 38
Table 12. Nutritional Se levels........ 42
Table 13. Suspect levels of various constituents of soils and overburden........ 68
Table 14. Percent of precipitation lost as runoff at five Northern Great Plains mine sites........ 76
Table 15. Common fertilizer forms of N........ 80
Table 16. Generalized grassland mineral N levels........ 81
Table 17. Soil P levels........ 83
Table 18. Common fertilizer forms of P........ 83
Table 19. Soil K levels........ 84
Table 20. Common fertilizer forms of K........ 84
Table 21. C:N ratios for some common materials........ 89
Table 22. Federal recommendations for maximum trace element loadings for agricultural lands........ 97
Table 23. Seed mix for a hypothetical Northern Great Plains coal minesoil........ 166
Table 24. Two seed mixes for a hypothetical topsoiled waste rock dump in the semiarid West........ 174
Table 25. Two seed mixes for a hypothetical waste rock dump without topsoil in semiarid western North America........ 175
Table 26. Seed mix for a coarse-textured coversoil on a hypothetical tailings pond in the semiarid West........ 187
Table 27. Seed mix for a hypothetical bentonite mine in the Black Hills mining region........ 194
Table 28. Seed mix for a hypothetical sand and gravel mine in the Northern Great Plains........ 197

Table 29. Seed mixes for an east-west highway at middle elevations in the Northern Great Plains 200
Table 30. Chemical constituents in coal ash 201
Table 31. Seed mix for a hypothetical ash disposal site on the Northern Great Plains 203
Table 32. Plant species used in eastern revegetation programs 213
Table 33. Plant species used in midwestern revegetation programs 217
Table 34. Plant species used in Northern Great Plains revegetation programs 219
Table 35. Plant species used in southwestern revegetation programs 223

LIST OF FIGURES

Figure 1. Soil textural triangle 11
Figure 2. Generalized water-soluble aluminum toxicity curve 27
Figure 3. Recontoured minesoil and adjacent native rangeland 60
Figure 4. Slopes, angles, and grades 63
Figure 5. Plowing a minesoil with an enlarged plow 70
Figure 6. A chisel on a minesoil 71
Figure 7. Agricultural disk at a coal mine 71
Figure 8. Chisel and harrow in final pass over a minesoil prior to drill seeding 72
Figure 9. A ripper at a coal mine 73
Figure 10. Gouged depressions on a minesoil 74
Figure 11. Germinating vegetation on a gouged minesoil 74
Figure 12. Cross sections of gouged depressions and dozer basins 75
Figure 13. Bulldozer attachment for creating catchment basins on a minesoil 75
Figure 14. Water saturated minesoil zones 77
Figure 15. First growing season vegetation on gouged minesoil 78
Figure 16. Plant response after three growing seasons on a gouged minesoil 78
Figure 17. Plant response to nutrient concentrations 80
Figure 18. Stylized nitrogen cycle 82
Figure 19. Manure spreader 86
Figure 20. Hydromulcher 86
Figure 21. Crimped straw 87
Figure 22. Hydromulch spray 87
Figure 23. Crimping straw 91
Figure 24. Two examples of erosion control blankets 93
Figure 25. Woven jute erosion control blanket 93
Figure 26. Plowing an amendment (lime) into an acid minesoil 99
Figure 27. Sprinkler irrigation on mined lands 100
Figure 28. Drill seeder 119
Figure 29. Cultipacker behind a disk 120
Figure 30. Drill-seeded plant community on a minesoil 121
Figure 31. Broadcast seeder or fertilizer spreader 122

Figure 32. Brillion broadcast seeder 123
Figure 33. Broadcast-seeded plant community on a minesoil 123
Figure 34. Seed dispersal by fixed wing aircraft 124
Figure 35. An example of containerized stock 129
Figure 36. Tree spade with wilding 131
Figure 37. Mechanized tree or shrub planter 132
Figure 38. Schematic representation of a shrub in a condensation trap 134
Figure 39. Two plants in a condensation trap 134
Figure 40. Sodded highway drainage 140
Figure 41. Plant propagules in direct-haul topsoil 146
Figure 42. The change of plant communities across a natural landscape in western North America 149
Figure 43. Topsoil depth variation and vegetation response on a natural hill slope in an arid or semiarid area 150
Figure 44. Impoundment on a western surface mine 153
Figure 45. Vegetation influenced by impounded water on a semiarid disturbed landscape 154
Figure 46. Final pit impoundment with intact highwall 155
Figure 47. Rock ledge constructed on a coal surface mine 158
Figure 48. Western coal stripmine 162
Figure 49. Open-pit metal mine 168
Figure 50. Waste rock at a western gold mine 171
Figure 51. Tailings being sluiced to a holding pond 179
Figure 52. Dry, light-colored tailings 183
Figure 53. Phosphate deposits of southeastern Idaho 189
Figure 54. Bentonite deposit ready for removal from the pit 192
Figure 55. Gravel pit with a high water table 196
Figure 56. Modern coal ash disposal pond with heavy membrane liner 202
Figure 57. Abandoned placer mine waste 205
Figure 58. Abandoned coal minespoils in the early 1970s 206
Figure 59. Abandoned wastes from hard rock mining and milling 206

INTRODUCTION

Four words occur with increasing frequency in the field of disturbed land revegetation. They are restoration, reclamation, rehabilitation, and revegetation.[1] Many people view these concepts as synonymous, but this author prefers to distinguish between them. In this text, the term *restoration* will be used to refer to the return of a degraded site to the exact ecological condition it exhibited prior to disturbance. *Reclamation* will be used to refer to the construction of topographic, soil, and plant conditions after disturbance, which may not be identical to the predisturbance site, but which permits the degraded land mass to function adequately in the ecosystem of which it was and is a part. *Rehabilitation* will be used as if it is synonymous with reclamation. The term *revegetation* is restricted to the vegetation phase of reclamation.

The distinction between these concepts becomes blurred in many articles and the differences may be relative. One may view restoration as the most concentrated effort to correct the problems introduced by degradation of the site. Reclamation may be considered a less intense and less expensive effort, and revegetation as an attempt to manipulate the surface plant covering only. It is convenient, however, to have a term to apply to high-intensity reconstruction efforts on landscapes. Efforts to duplicate tall grass prairie in Iowa on land that had been plowed for a cornfield many years ago, or attempts to create wetlands on muck soils drained for crop production, are examples of what is referred to here as restoration. These types of activities are distinct from the efforts by mining companies, highway departments, or drilling contractors, who emphasize ecosystem integrity, geographic and soil stability, and native vegetation in applied reclamation.

Rehabilitation of degraded sites will occur with no input from man if sufficient time is allowed for the necessary edaphic and biologic changes to take place.[2] These changes are controlled by the type of rootzone material on the site at the start of the stabilization process, the climatic regime of the area, and the accidental dissemination of plant propagules to the disturbed soil. While adequate nutrient concentrations accumulate in degraded soils of mesic regions in as little as 50 years,[3] soil changes are insignificant in semiarid minesoils in that same amount of time.[4]

If man does not intervene, a disturbance is rehabilitated in a series of stages called plant succession. Revegetation and soil stabilization begin immediately after any perturbation when nontoxic rootzone materials remain after disturbance. The first stage of plant succession consists of the invasion and germination of representatives of a type of plant referred to as pioneers or opportunists. These plants, which are characterized by large numbers of easily distributed seeds (probably airborne), are often annuals or biennials. They exhibit tolerances for temperature extremes, elemental imbalances, site disturbance, nutrient-poor soils, and drought. At the same time, they have high light requirements and rarely do well in shady environments. As these plants mature, they stabilize the soil on the disturbance, reduce surface soil temperatures, increase soil organic matter, and contribute to the development of nutrient pools. The pioneer species create habitat characteristics that are different from those present on the disturbance when they germinated.

These early successional plant species change the soil and the environment immediately above and below the surface. The soil is now more erosion resistant because of the maze of plant roots it contains and because of the ground cover of plant tissues. These leaves shade the soil surface, thereby lowering its temperature and reducing water loss through evaporation. The soil characteristics are modified enough that the site is now capable of supporting a broader range of plant species — plant species that will not tolerate the extreme soil, nutrient, and moisture conditions of the original disturbance.

Later successional or seral species begin to crowd into the plant community on the aging disturbance. These plants are longer lived than most opportunists. They begin to accumulate soil nutrients and push their roots into spaces occupied by the older and dying pioneer species. The seral species continue the site modification initiated by the earlier arrivals. They so modify the habitat that most pioneers cannot successfully reproduce on the site. Subsequently, additional plant species continue the change in vegetation on the formerly disturbed site. The perpetuation of these changes in plant species is called plant community succession.

Each successive wave, group of plants, or seral community lasts longer then its predecessor. Eventually, habitat changes and the subsequent movement of new plants into the community cannot be measured in a reasonable length of time. At this point, ecologists say climax has been reached and the plant community is at its point of maximum stability. Unfortunately, hundreds to thousands of years are required for the plant community to reach this stage of succession in semiarid to arid regions.

In terms of the demands placed on all lands today, the slowness of the natural reclamation process is not acceptable. The needs of modern society necessitate that disturbed areas be returned to some type of stable ecosystem (not actively eroding) as rapidly as possible. Therefore, the responsibility of the reclamation scientist is to condense the successional process on degraded soils into a few years. This text will examine numerous relationships, principles, and techniques that the revegetation scientist may employ to accelerate the successional process so that a stable, and hopefully productive, plant community develops on the disturbed site in a relatively short time.

This text has been prepared for the land reclamation student or rehabilitation professional. It emphasizes the revegetation of disturbed lands from the vantage point of the plant specialist. It is not intended to cover the entire reclamation program, but only the final and most visible component of the plan: the development of vegetation on drastically disturbed soils. Discussion must start with the soil resource because it is the medium that supports the plant root. Therefore, the soil will be studied insofar as a disturbance may render it incapable of supporting green plants. Soil properties, especially as they directly impact plant performance, will be discussed.

Because rehabilitating disturbances in semiarid to arid regions is so difficult, most specific comments will be directed to revegetation in such ecologic regions. Examples will cover the spectrum of land disturbance from highway rights-of-way, through oil and gas drilling pads, power line corridors and pipelines, to minesoils, mine tailings, and hazardous inorganic waste sites.

Examples will illustrate principles and will be related to the rationale behind many reclamation requirements of state and federal regulatory agencies. Hopefully, this will help the rehabilitation specialist develop more efficient revegetation programs. To understand regulatory requirements and the revegetation process, however, it is necessary to review relationships between the physical and chemical environment of the soil-plant-animal system. Relationships between soils and the plant community will be reviewed in the first sections of this text. Soil-plant-animal interactions will be emphasized in the middle chapters of the book. The integration of soils and plants into complete revegetation efforts will complete the text.

Seed germination, seedling establishment, plant growth, and plant reproduction are a series of events that depend on a host of soil and climatic factors, as well as past and present plant physiologic activity. We will speak of each of these activities as if they were independent phenomena, but on any landscape they are interrelated, overlap, and are intimately interdependent. In this text we will discuss these plant phenomena as they relate to the soil-plant-animal system developing on the disturbed site. We will discuss the relationship of ecologic processes and the different stages of the plant life cycle. However, the student of the land rehabilitation process must be cognizant that, in the field, these life stages cannot be separated. Statements made about one segment of a plant life cycle will apply to other stages of plant development.

Enactment of the Surface Mining Control and Reclamation Act (SMCRA) in 1977 determined the direction that all revegetation plans for coal mines would take in the United States. Some state mine regulatory agencies have used this program, with modifications, as a model for the regulation of revegetation programs for disturbances other than those produced by coal mining. Provisions similar to those promulgated by SMCRA are often applied to metal mining, pipeline corridors, transmission rights-of-way, and other disturbances. Therefore, this legislation will be referred to frequently in this text and cited as a model of one of the better mined-land rehabilitation laws in this country's history.

REFERENCES

1. Box, T. W., The significance and responsibility of rehabilitating drastically disturbed lands, in *Reclamation of Drastically Disturbed Lands*, Schaller, F. W. and Sutton, P., Eds., American Society of Agronomy, Madison, WI, 1978, 1.
2. Johnson, M. S. and Bradshaw, A. D., Ecological principles for the restoration of disturbed and degraded lands, *Appl. Biol.*, 4, 141, 1979.
3. Leisman, G. A., A vegetation and soil chronosequence on the Mesabi iron range spoil banks, Minnesota, *Ecol. Monogr.*, 27, 221, 1957.
4. Schafer, W. M., Nielson, G. A., Dollhopf, D. J., and Temple, K., *Soil Genesis, Hydrological Properties, Root Characteristics, and Microbial Activities of 1 to 50-Year Old Stripmine Spoils*, U.S.E.P.A. 600/7-79-100, U.S. Environmental Protection Agency, Cincinnati, OH, 1979.

1

Soil Parameters That Directly Affect Plant Establishment and Development on Disturbed Lands

Revegetation of any disturbance depends on an understanding of the chemical and physical properties of the material in which the plants will be rooted. Therefore, regulatory agencies usually require that numerous soil parameters be determined before the site is disturbed. Ostensibly, because soil chemical or physical characteristics affect plant performance or ground water quality, they must be known before the site reclamation plan can be completed. Certain soil parameters are particularly important to the revegetation scientist. Those rootzone characteristics considered most important may vary in different ecologic regions, but several are common to analyses required almost worldwide. These major soil parameters and a few additional soil characteristics of importance in semiarid or arid regions are briefly reviewed in this chapter. In addition, mention will be made of relevant analytical procedures for determining each parameter. For a more thorough evaluation of the soil resource, the reader may consult any reputable soil textbook.

HYDROGEN ION CONCENTRATION

One critical soil characteristics is the hydrogen (H) ion concentration. This parameter is normally recorded as the pH or the negative log of the hydrogen ion concentration. Soil pH is important because it influences many other soil chemical properties. For example, the availability of biologically active elements to the plant varies considerably with soil pH. At values below 5.5, metal levels may become so high and phosphorus levels so low that they limit plant growth. At pH values above 8.0, some plant nutrients appear to be deficient for plant needs. Generalized soil pH ranges are shown in Table 1.

While a soil solution pH of 5.5 indicates a strongly acid soil, it is not uncommon to find disturbed lands revealing pH values even

TABLE 1 Soil Reaction Ranges[1]

pH Range	Description
<4.5	Extremely acid
4.5–5.0	Very strongly acid
5.1–5.5	Strongly acid
5.6–6.0	Moderately acid
6.1–6.5	Slightly acid
6.6–7.3	Neutral
7.4–7.8	Slightly alkaline
7.9–8.4	Moderately alkaline
8.5–9.0	Strongly alkaline
>9.0	Very strongly alkaline

lower. Many derelict and active mine sites — both coal and hardrock mines from Canada to Mexico and from the western prairies and mountains to the eastern coal fields — reveal surficial soil materials with values below pH 4. Such high hydrogen ion concentrations release large quantities of metals into the soil solution. At these low pH levels, it is questionable whether the hydrogen ion concentration or elevated metal levels are responsible for the death of vegetation.

Soil pH values are normally determined in a saturation paste or in the extract prepared from a soil paste. A review of this method and other methods of pH analysis may be found in the U.S. Department of Agriculture, Handbook No. 60.[2]

Vegetation performance is usually maximum in the neutral pH range (6.6 to 7.3), but values from slightly acid to slightly alkaline are common in natural environments. In extreme instances, tolerant plant species may be found performing adequately in soils ranging from moderately acid to moderately alkaline pH unless other chemical or physical factors inhibit plant performance.

ELECTRICAL CONDUCTIVITY

Salinity is a soil property referring to the amount of soluble salt in the soil. It is generally a problem of arid or semiarid regions. Electrical conductivity (EC) is the most common measure of soil salinity (Table 2). Soil EC values may be determined in the aqueous soil extract used for the determination of soil pH.[4] By agricultural standards, soils with an EC greater than 4 dS/m (decisiemens/meter) are considered saline.[5]

TABLE 2 Soil Salinity Guide[3]

Parameter	Nonsaline	Slightly Saline	Moderately Saline	Saline
EC (dS/m)	<4	4–8	8–16	>16

Actually, salt-sensitive plants may be affected at conductivities less than 4 dS/m and salt tolerant species may not be impacted by concentrations of up to twice this maximum agricultural tolerance limit. Thus, the reclamation scientist must exercise care in the interpretation of salinity standards. Large areas of western rangelands that are saline by these standards support productive native grasslands. Salinity, therefore, should be defined in terms of the predisturbance land use potential, the proposed postdisturbance land use, and the plant species to be seeded on the site.

SODIUM IN SOILS

Sodium (Na) in soils contributes to their alkalinity, as opposed to salinity which is caused by any type of salt. Elevated concentrations of sodium ions create a plant growth hazard which is measured by one of two methods. The more common technique, the sodium absorption ratio (SAR), is the proportion of sodium ions compared to the concentration of calcium (Ca) plus magnesium (Mg) ions in the saturation paste. The second method of estimating the sodium hazard is called the exchangeable sodium percentage (ESP). The ESP refers to the concentration of sodium ions on cation exchange sites.

The SAR is defined as

$$SAR = \frac{Na}{\sqrt{(Ca + Mg)/2}} \qquad [1]$$

The concentrations of sodium, calcium, and magnesium are expressed as milliequivalents per liter (meq/l) in the solution extracted from a saturated soil paste.[6] When the SAR rises above 12 to 15, serious physical soil problems arise and plants have difficulty absorbing water.

The ESP is defined as the percent of the total cation exchange dominated by sodium,[7] or

$$ESP = \frac{Na}{\text{Cation Exchange Capacity}} \times 100 \qquad [2]$$

All concentrations are meq/100 g of soil as determined in an ammonium acetate extract.

In general, ESP values are slightly higher than SARs at the levels found in better rangeland soils. A high SAR (>12 to 15) or ESP (>15%) indicates that sodium concentrations in the soil are too high. Such soils are referred to as sodic or alkaline (Table 3). There is no universal agreement as to the inhibitory properties of these elevated SAR or ESP values since many productive western rangeland soils reveal sodium concentrations that place them in the sodic category. Salinity,

TABLE 3 Suggested ESP and SAR Ranges

Rank	ESP	
Nonalkaline	<15–<10	Assumes heavy clay soil, if clay content is <20% these values may be increased up to 25%
Moderately alkaline	10–15	
Alkaline	>15	
Rank	**SAR**	
Adequate	<11	Assumes heavy clay soil, if clay content is <20% these values may be increased up to 25%
Fair	11–15	
Poor	>15	

texture, and clay mineralogy exert strong interactions with, and directly influence, sodium impact on plant performance. These influences prevent the development of a clear and concise statement of the effect of sodium on plants. The values in Table 3 may not be appropriate for mined lands because of these other factors (i.e., clay and salinity); nevertheless, these values may serve as a guide for disturbed sites as well as native range soils.

ORGANIC MATTER

Soil organic matter is often defined as the percent of humus in the soil. Humus is the unidentifiable residue of plants and animals that has become fairly resistant to further decay. Reclamationists, however, use the term soil organic matter in a more general sense. They include in the term any carbon assemblages such as straw, manure, wood chips, mulches, etc., that are incorporated into the soil.[8] From this point of view, soil organic matter is composed of two fractions: (1) the recognizable organic material (straw, woodships, roots, insects, etc.) with their partially decomposed remains, and (2) humus. This broader concept of organic matter is commonly used in discussions about soil rehabilitation. Typical Great Plains topsoils reveal organic matter contents in the very low to low range (Table 4); however, the low levels of rangeland organic matter do not negate its importance. Organic matter by any definition increases soil porosity, infiltration, CEC (cation exchange capacity), water-holding capacity, and nutrient reserves. It also improves soil structure and lowers bulk density.

TABLE 4 Soil Organic Matter Levels[9]

Rank	Soil Organic Matter (%)
Very low	<2.0
Low	2.1–3.5
Medium	3.6–5.0
High	5.1–10.0
Very high	>10.0

Obviously, organic matter is very important to the functioning soil system.

Important steps in the rehabilitation of functioning nutrient cycles involve: (1) the uptake of essential nutrients (especially nitrogen [N], phosphorus [P], potassium [K], and sulfur [S]) by plants; (2) the release of these nutrients by decomposing organic material; and (3) the maintenance of these nutrients in a pool available to plants. An understanding of how organic matter contributes to these soil functions is extremely important to the revegetation scientist.

Organic matter contains small quantities of nitrogen, phosphorus, and other plant nutrients. It also forms an energy source (carbon [C] compounds) for many decomposers. Bacteria, fungi, and other soil decomposer organisms break complex organic molecules into simpler elemental forms. This process releases energy in the compounds as fuel for the organism's life processes. As microorganisms break carbon chains into successively smaller pieces they release plant nutrients. These different ions or elemental forms of nitrogen, phosphorus, or other essential elements are incorporated into the microorganism protein or released into the soil solution. In the soil solution these nutrients may be absorbed by plants, leached from the soil, attached to cation exchange sites, or precipitated by biological or chemical processes. Carbon is ultimately released to the atmosphere as carbon dioxide (CO_2).

The revegetation scientist must be careful to distinguish between organic matter and organic carbon. As a general rule, soil organic carbon contents are determined by techniques such as the Walkley Black method. In this method, the organic carbon figure is multiplied by some value (1.7 to 2.0) to generate an estimate of the amount of organic matter in the soil. Determining the amount of organic matter in disturbed soils of dry regions is difficult because inorganic carbon (as carbonates) is present in most of these soils. Under these conditions, we may determine the concentration of organic carbon by analyzing inorganic carbon and total carbon and subtracting inorganic carbon from the total.[10]

The ratio of soil organic matter to nitrogen is very important. It is often referred to as the carbon:nitrogen (C:N) ratio. A C:N ratio of 20:1 or lower is adequate to supply soil microbes and plant nitrogen needs. A ratio in this range indicates that some nitrogen released by the decomposition of organic matter is available to higher plants. Higher C:N ratios (e.g., 25:1) provide an adequate food and energy source for bacteria; however, the amount of nitrogen is so low that most of it is incorporated into bacterial protein and little is available for higher plants. At these higher ratios, nitrogen is immobilized in the protoplasm of soil microorganisms. Under these conditions the soil nitrogen supply continues to limit higher plant growth until bacteria and other organisms decompose organic matter, reduce the amount of carbon in the soil, and increase the soil nitrogen available to higher plants.

TEXTURE AND STRUCTURE

Texture refers to the relative proportion by weight of the soil fractions (sand, silt, and clay) in a sample. The soil aliquot used for this analysis should be composed of particles that pass a 2-mm sieve. The names of the textural classes reflect the relative distribution of the three primary particles in the soil. Thus, sand refers to a soil composed predominately of sand-sized particles (Table 5). Sandy clay refers to a soil that is composed of 35 to 55% clay, 45 to 65% sand, and/or 0 to 20% silt (Figure 1). Any soil composed of all three particle sizes and not exhibiting the physical properties of any one of them is referred to as a loam. Texture influences plant growth by its direct effect on soil aeration, water infiltration, CEC, and erodibility. Infiltration and permeability are rapid in sandy soils, very slow in clay soils, and intermediate in loam soils. Soil texture can be estimated in the field by experienced soil scientists. It can be measured in the laboratory by the pipette or hydrometer methods.[12]

The aggregation of primary soil particles into larger fragments contributes to a soil property referred to as soil structure. This soil characteristic varies from granular to massive. Granular soils have many large and small pores — a desirable characteristic for plant growth. Soils described as massive have few large pores and exhibit poor water infiltration.

BULK DENSITY AND POROSITY

The weight of a unit volume of a soil structural component — including its pore space, organic matter, and soil particles — is referred to as the bulk density of the soil. Bulk densities range from less than 1.0 g/cm^3 to greater than 2.0 g/cm^3 (Table 6). In general, sands have higher bulk densities (1.3 to 1.7) than finer textured soils

TABLE 5 Diameter Range of the Primary Soil Particles[11]

Particle	Diameter (mm)
Clay	<0.002
Silt	0.002–0.05
Sand	0.05–2

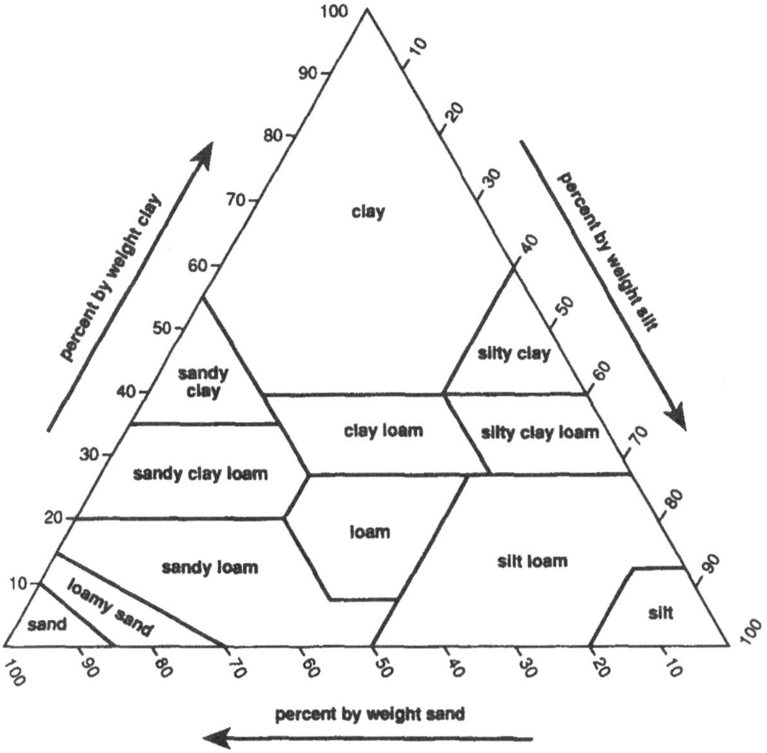

Figure 1 Soil textural triangle.

TABLE 6 The Range of Soil Bulk Density and Porosity

Bulk Density (g/cm³)	Rank	Porosity (%)
<1.3	Low	<30
1.3–1.7	Medium	30–60
>1.8	High	>60

(1.1 to 1.6).[13] Soils have different bulk densities within a textural class or over time because of a change of structure. This may be caused by machinery passing over the soil, which decreases pore volume, or by an addition of organic matter, which increases pore volume. One problem of disturbed site revegetation is the increase in soil density caused by vehicular compaction of the soils during the rehabilitation process. Increasing soil compaction restricts the permeability of a soil to water and hinders root penetration.

Soil porosity refers to that part of a soil volume that is not occupied by soil particles or organic matter. Pore spaces are filled with either air, other gases, or water. Large pores (macropores) are filled with air or serve as passages for water. They make the soil permeable to air and water and provide channels for root growth.[14] High permeabilities permit faster infiltration and percolation of water through a soil or soil horizon. Small pores (micropores) exhibit attractive forces strong enough to hold water in the pore. They are the water retention system of the soil. During precipitation, macropores conduct water into the soil where it fills the micropores. At field capacity all pores small enough to retain water against the pull of gravity are filled.

Clay soils have numerous micropores and hold large quantities of water, but since they have few macropores they produce very slow infiltration rates. The pores in the clays may be so small and hold water so tenaciously that the water is not available to plants. Sandy soils with numerous macropores but few micropores have higher infiltration and percolation rates (Table 7) but a lower water-holding capacity than other soil textures. A soil with a stable crumb structure may exhibit 60 to 70% porosity that is equally distributed between micro- and macropores. A clay soil may have a porosity almost equal to that of a soil with a crumb structure, but the pores in the clay soil are composed predominately of micropores. A silty or clayey soil that has been compacted by reclamation equipment may still exhibit 40% or more porosity, but as previously noted it may have no macropores.[16] Therefore, regulatory restrictions on the use of textural extremes (sands or clays) in soil reconstruction are very appropriate. The best performance of most desirable plants is expressed in intermediate soil textures (loams) where the soils contain mixtures of micro- and macropores.

CATION EXCHANGE CAPACITY

The quantity of cations that can be adsorbed by a soil is commonly referred to as its cation exchange capacity (CEC). Cations (e.g., Ca^{2+}, Mg^{2+}, Na^{+}, K^{+}, H^{+}) attracted to and adsorbed on negatively charged

TABLE 7 Soil Permeability Rates[15]

Rank	Permeability (mm/h)	Generalized Texture
Very slow	<1.5	Fine
Slow	1.5–5	
Moderately slow	6–15	
Moderate	16–50	
Moderately rapid	51–150	
Rapid	151–500	
Very rapid	>500	Coarse

sites on soil colloids are said to be on the cation exchange. They are in equilibrium with cations in the soil solution, but resist loss by leaching because of their attraction to soil colloids. Anions (e.g., PO_4^{2-}, MoO_4^{2-}, SO_4^{2-}) are present in the soil solution and may be weakly attracted to exchange sites saturated with cations. When ions in the soil solution are depleted by crop removal or leaching, adsorbed cations or anions exchange with other ions that may be in temporary excess in the soil solution. This action maintains a relatively constant soil solution nutrient pool. The long-term loss of ions from the soil solution and exchange sites is balanced by the decomposition of soil organic matter and weathering of soil parent materials. Problems develop, however, when certain ions are removed from the soil faster than they can be supplied by weathering or decomposition. Thus, in agricultural soils in which crop removal depletes the nutrient ion supply, or in soils influenced by the oxidation of iron pyrite (FeS_2), hydrogen ions become more numerous on the exchange. Such soils become acidic when the rate of hydrogen ion accumulation exceeds the rate of hydrogen ion removal.

A rough relationship exists between soil texture and CEC. In general, finer textured soils reveal higher exchange capacities and coarse textured soils have lower exchange capacities (Table 8). Some clay soils have CECs as high as 150 meq/100 g of soil and humus may have CECs on the order of 300 meq/100 g of material.

OTHER SOIL TESTS

Many other chemical or physical characteristics of soils can be determined (e.g., oxidation-reduction potential, mineralogy, buffering capacity, and B concentration). If these properties are important at certain reclamation projects the rehabilitation scientist should determine them at the particular site. For example, the oxidation-reduction potential of a soil is important at metal-contaminated sites and the buffering capacity of spoil or soil is important in acid-generating materials. These tests for specific local problems are not always necessary for the revegetation of disturbed lands.

TABLE 8 General Ranges of Soil CEC[17]

Rank	CEC (meq/100 g)	Type of Material
Low	<5	Sand
Medium	5–30	Loams
High	>30	Clay
	125–400	Peat

REFERENCES

1. Soil Survey Staff, *Soil Survey Manual, Handbook No. 18,* U.S. Department of Agriculture, Washington, D.C., 1951, 235.
2. Richards, L. A., *Diagnosis and Improvement of Saline and Alkaline Soils,* Agriculture Research Service Handbook No. 60., U.S. Department of Agriculture, Washington, D.C., 1969, 17.
3. Soil Conservation Service, *National Soils Handbook,* U.S. Department of Agriculture, Washington, D.C., 1983, 603–26.
4. Rhoades, J. D., Soluble salts, in *Methods of Soil Analysis, Part 2,* 2nd ed., Page, A. L., Ed., American Society of Agronomy and Soil Science Society of America, Madison, WI, 1982, 167.
5. Richards, L. A., *Diagnosis and Improvement of Saline and Alkaline Soils,* Agriculture Research Service Handbook No. 60, U.S. Department of Agriculture, Washington, D.C., 1969, 9.
6. Richards, L. A., *Diagnosis and Improvement of Saline and Alkali Soils,* Agriculture Research Service Handbook No. 60, U.S. Department of Agriculture, Washington, D.C., 1969, 26.
7. Carlstrom, M. G., Amendola, F. A., Shay, D. A., and Dollhopf, D. J., Sodium, in *Reclaiming Mine Soils and Overburden in the Western United States,* Williams, R. D. and Schuman, G. E., Eds., Soil and Water Conservation Society, Ankeny, IA, 1987, 75.
8. Smith, P. L., Redente, E. F., and Hooper, E., Soil organic matter, in *Reclaiming Mine Soils and Overburden in the Western United States,* Williams, R. D. and Schuman, G. E., Eds., Soil Conservation Society, Ankeny, IA, 1987, 185.
9. Energy Laboratories, Inc., *Soil Analysis Interpretation Guide,* Billings, MT.
10. Nelson, D. W. and Sommers, L. E., Total carbon, organic carbon, and organic matter, in *Methods of Soil Analysis, Part 2,* 2nd ed., Page, A. L., Ed., American Society of Agronomy and Soil Science Society of America, Madison, WI, 1982, 539.
11. Soil Conservation Service, *National Soils Handbook,* U.S. Department of Agriculture, Washington, D.C., 1983, 603–5.
12. Gee, W. G. and Bauder, J. W., Particle size analysis, in *Methods of Soil Analysis, Part 1,* 2nd ed., Klute, A., Ed., American Society of Agronomy and Soil Science Society of America, Madison, WI, 1986, 377.
13. Miller, C. E., Turk, L. M., and Foth, H. D., *Fundamentals of Soil Science,* 3rd ed., John Wiley & Sons, New York, 1958, 51.
14. Hamblin, A. P., The influence of soil structure on water movement, crop root growth, and water uptake, *Adv. Agron.,* 38, 95, 1985.
15. Soil Conservation Service, *National Soils Handbook,* U.S. Department of Agriculture, Washington, D.C., 1983, 603–19.
16. Bonneau, M. and Souchier, B., *Constituents and Properties of Soils,* Academic Press, New York, 1982, 274.
17. Donahue, R. L., Miller, R. W., and Shickluna, J. C., *Soils,* 5th ed., Prentice-Hall, Englewood Cliffs, NJ, 1983, 115.

2

Plant and Soil Elemental Chemistry

Plants are composed of almost every element known to man, but only a limited number of these elements are of particular importance to the plant. Some biologically active elements are essential for plant growth and development (e.g., C, B, H, O, N, P, K, Ca, Mg, S, and Fe); some are toxic to the plant (e.g., Al, As, and F). Of those elements that are essential for plant growth, carbon, hydrogen, and oxygen (O) are the main constituents of protoplasm. Plants acquire these elements from carbon dioxide (CO_2) that diffuses into plant tissues from the atmosphere and water (H_2O) that is usually absorbed from the soil via the plant root system. Other elements, both essential and nonessential, may be acquired by the plant as it absorbs water or, in a few instances, as gaseous forms that diffuse into the leaf with CO_2.

The chemical symbols and common ionic forms of the essential elements absorbed by plant roots are shown in Table 9. Small amounts of other elements may be necessary for an individual plant species to complete a life cycle (e.g., nickel [Ni]), but those shown in Table 9 represent the major plant nutrients. Most of these are also essential for normal animal growth and development. Plants absorb some elements not known to be required by them but which are necessary for animal nutrition (e.g., selenium [Se]). All of these elements, and several others known to be toxic to plants and/or animals, will be discussed in this chapter.

Elemental levels in soils are determined by a number of analytical techniques. The most common methods involve removal of the element or elements from the soil or overburden by an extracting liquid and the determination of the concentration of the particular element in the extracting solution. The method may be designed to extract one particular ion (e.g., NO_3^-, PO_4^{2-}) or the extractant may remove multiple ions or elements from the soil. Extractants designed to remove several elements simultaneously include water (especially hot water), ammonium acetate, diethylenetriaminepentaacetic acid (DTPA), ammonium bicarbonate-DTPA (AB-DTPA), and other chelating agents,

TABLE 9 Plant Nutrients[1]

Nutrients	Chemical Symbol	Ionic Forms Commonly Absorbed by Plants
Macronutrients		
Nitrogen	N	NO_3^-, NH_4^+
Phosphorus	P	$H_2PO_4^-$, HPO_4^{2-}
Potassium	K	K^+
Calcium	Ca	Ca^{2+}
Magnesium	Mg	Mg^{2+}
Sulfur	S	SO_4^{2-}
Micronutrients		
Manganese	Mn	Mn^{2+}
Iron	Fe	Fe^{2+}, Fe^{3+}
Zinc	Zn	Zn^{2+}
Copper	Cu	Cu^{2+}
Molybdenum	Mo	MoO^{2-}
Chlorine	Cl	Cl^{4-}
Boron	B	BO_3^{3-}

acids, and bases. Most of these techniques were designed to determine elemental deficiencies in agricultural soils. The use of these methods to determine elemental levels in minesoils, overburden, or violently disturbed soils is of questionable validity. Soil extractants designed specifically for the determination of toxic elemental levels in violently disturbed soils or overburden have not been developed. Because knowledge about extractants for these materials is lacking, the rehabilitation scientist is forced to rely on existing agricultural soil extraction techniques.

Whenever reclamation specialists describe soil elemental levels in the preparation of reports, in applications for various permits, or in the preparation of legal documents, it is extremely important that the method of extraction and determination be stated. For example, total content of an element in the soil is many times plant-available, extractable, or soluble soil contents. These fractional pools may differ from one another by very large amounts. While some agencies, laboratories, or mine companies require that very specific techniques be used to remove an element from the soil, the methodology is less important than the knowledge of a relationship between the amount extracted by a particular technique and plant response.

Information about plant response and elemental levels determined by one extraction technique cannot be extrapolated to other methods. For example, the amounts of zinc extracted by ammonium acetate are not comparable to amounts extracted by DTPA or water. In a similar manner, the amount of boron found in the extract from a soil paste cannot be equated to hot-water-soluble boron and can only be weakly related to displaced soil solution boron. Unless elemental levels can be related to plant or animal performance the exact analytical tool used for soil analysis is of minor importance and numbers representing the concentration of any element or ion in an extractant solution are meaningless.

Soil analyses enable us to look at the soil at one moment in time, but the rehabilitation scientist is concerned with long-term reclamation of disturbed sites. The brief picture we acquire tells us nothing of long-term changes in the soil. In the past, it has been assumed that the status of soils on rehabilitated sites remained relatively unchanged after reclamation; however, we now know that rapid changes take place in minesoils after reconstruction. Significant soil changes have been recorded in less than 50 years on minesoils developing on midwestern mine wastes.[2] These changes may improve the soil or they may contribute to degradation of the resource. One of the changes anticipated in minesoils is the oxidation of reduced compounds in overburden brought to the near-surface environment. Two elements associated with changes induced by oxidation of near-surface materials are selenium and sulfur.

The acidification of surface soils on successfully revegetated minesoils has occurred several years after completion of reclamation. Acid soils are a product of the oxidation of reduced sulfur compounds brought to the near-surface environment by mining. This action indicates that the status of a soil on a disturbed landscape cannot be maintained, and that changes will occur in this resource after rehabilitation. In semiarid and arid regions, the oxidation of selenium from reduced forms found in buried horizons to water-soluble plant-available forms in the surface soil environment may lead to future selenium excesses in disturbed land vegetation.

The reclamation scientist must determine whether changes in degraded soils after revegetation will be within acceptable limits of plant growth and animal performance. Brief studies that appear to support physical or chemical soil properties as noninhibitory to vegetation or animal performance are not adequate for such purposes because soil properties will change several years after rehabilitation. Until the time when the chemical properties of spoils and disturbed soils are more clearly understood, those responsible for revegetation of degraded lands must use the best available technology and estimates of what rootzone materials will produce acceptable plant performance over many years of plant growth.

MACRONUTRIENTS

Some biologically active elements are required for plant growth in such large quantities that they are referred to as macronutrients. These include nitrogen, phosphorus, potassium, calcium, magnesium, and sulfur. Iron is needed in smaller quantities but is included among the macronutrients by some botanists. The group of biologically active elements effective in minute quantities are known as trace elements. If the trace element is required for plant growth it is called a micronutrient. Iron is included with the micronutrients in this text.

Nitrogen

The common plant-available form of this element is the nitrate ion (NO_3^-). This ion is the chemical constituent extracted from the soil by plants that is most likely to be limiting to plant growth. Nitrogen, derived from this ion is an integral component of all proteins, enzymes, and other biologically active substances in both plants and animals. Nitrogen deficiency produces small pale plants and retards physiological development. Typical plant levels of this element range from 2 to 4% in high-yielding legumes such as alfalfa *(Medicago sativa)* and from 0.25 to almost 2% in most range grasses.[3] The nitrogen content of plants varies widely as a function of species, soil nutrient levels, and age of the tissue. Younger, physiologically more active tissues show higher nitrogen levels than older, less active tissues.[4]

Much of the nitrogen reserve is stored in the soil as organic matter and most of this organic fraction is found in the upper soil horizons. At surface mines and other types of surficial disturbance (e.g., highway shoulders and pipeline corridors) the upper soil horizons are usually removed and stockpiled prior to disturbance. Stored topsoil reveals a relatively rapid conversion of organic nitrogen to soluble nitrate[5,6] and possibly conversion of nitrate to ammonium ions (NH_4^+) deeper in the stockpile.[7] While these ions may be absorbed and utilized by most plants, they are also subject to volatilization, fixation by clays, or conversion back to nitrate. On the other hand, nitrate ions are soluble and subject to leaching or conversion to free nitrogen (denitrification) and volatilization. Thus, when stored topsoil is spread on a disturbed landscape, nitrogen reserves may have been depleted or altered by several chemical or biological phenomena and the healthy cycling of nitrogen through the ecosystem inhibited or prevented.

The establishment of a nitrogen cycle in disturbed soils is very important (for a discussion of this topic see *Nitrogen Fertilization* in Chapter 3). An active soil nitrate pool has little impact on seed germination or plant establishment but has pronounced effects on seedling growth.[8] In soils with adequate organic matter concentrations, the nitrogen pool is constantly replenished by decomposition of this reservoir. However, if the soil organic matter pool is small the cycling of nitrogen will be impeded. In degraded soils without topsoil or in poor quality stored topsoil, organic contents and nitrogen levels are reduced. If large quantities of organic materials are added to the soil the regrowth of microbial populations is very rapid. These organisms complex all available nitrogen before any excess is available for higher plant use. Therefore, in poor quality rootzone materials, plants find an inadequate supply of nitrogen, growth rates are reduced, and species composition of the plant community changes. Fertilizers, seeds of nitrogen-fixing plant species, and organic matter are often added to disturbed lands to augment supplies of critical components of the nitrogen cycle and to encourage rapid reestablishment of a normally functioning cycle in the newly constructed soil.

Soil nitrogen may assume a variety of forms. It is very dynamic and is constantly changing chemical species and concentrations. In most soils nitrate is the common ionic form of plant-available nitrogen, but this element may also exist as ammonium or nitrite (NO_2^-) as well as other ions. Nitrate concentrations vary in the soil as a function of season, plant growth rates, climatic conditions, and amount and type of vegetation on the site. In addition, microbes are continually breaking down organic matter and turning over themselves (dying and decomposing) — both processes releasing nitrogen in various forms into the soil solution.

Despite the obstacles introduced by the complex interactions of this element, it is often desirable to estimate the static concentration of plant-available nitrogen in the soil. In the western plains, the major portion of the soil nitrogen reserve is in the form of nitrates or nitrites; therefore, the potassium chloride (KCl) extraction, cadmium (Cd) reduction, and colorimetric determination of nitrates has been recommended for static soil nitrogen determinations in this region.[9] Nitrogen soil levels determined by this technique often are below 30 kg/ha (27 lb/acre). While low, these levels are adequate for the growth of most native range plants. Other methods have also been utilized to determine nitrogen concentrations because of the poor correlation between soil nitrate concentrations and plant response. Total nitrogen has been determined by the Kjeldahl procedure and nitrogen mineralization potential determined by various incubation techniques. None of these techniques provides a better estimate of soil nitrogen and plant growth relationships than nitrate analysis. Texts containing descriptions/discussions of disturbed land nitrogen concentrations must include a statement about the analytical procedure used to extract and to determine the levels reported.

Elevated nitrates are not a problem in native soils and plant toxic nitrate concentrations in disturbed soils do not occur. If the soil nitrate concentration exceeds 10 μg/g, however, it is probable that some leaching will take place in mesic regions. It is unlikely that leaching of nitrogen occurs in the West even at this soil level because of the lack of precipitation. Elevated nitrate concentrations in soils and ground water have been reported at some mines,[10] but it was not clear if this nitrate was derived from contaminated overburden or leaching.

Phosphorus

Phosphorus plays an extensive role in plant and animal metabolism. It is usually available in soil as the phosphate ion (PO_4^{2-}). This element is particularly important during seedling growth, and therefore is especially important to the plant shortly after seed germination. Plant phosphorus levels may range from 0.2 to 0.3% in alfalfa and from less than 0.1 to almost 0.3% in most grasses.[11] As with plant nitrogen levels, phosphorus concentrations in plant tissues are strongly influenced by tissue age.[12] This is a reflection of the physiologic activity of the plant tissue. Younger tissues are more active and reveal higher phosphorus concentrations. Deficiency symptoms are often expressed by reduced growth in seedlings, reddish-purple discoloration of the foliage or stems, and possibly death of the leaf tips.

The major portion of the total soil phosphorus reserve — 96 to 99% in semiarid rangeland soils — is not plant-available. It is complexed in organic matter, by metals in acid soils, or by calcium at higher pH values. On disturbed soils in the West, phosphorus extract concentrations are usually low to medium, but if soil organic matter levels are adequate a continuing supply of phosphorus becomes available for plants as the organic material decomposes. Typical available phosphorus levels in western soils range from <5 to >10 μg/g (Olsen extraction).[13] It is unclear what the range of concentrations determined by the Bray extraction would reveal because of the interference of excess calcium carbonate with this extraction. The recently developed AB-DTPA extraction technique removes soil phosphorus at concentrations approximately one half of the ranges reported for the Olsen extraction.[14] Because of the differences in results from extraction techniques, it is important that the reclamation specialist know the exact method used in the laboratory analyzing the samples. In western soils with high carbonate concentrations, the Olsen or bicarbonate extraction technique and colorimetric determination of phosphorus have been recommended.[15] In more mesic regions with low calcium carbonate soils, the Bray extraction technique is commonly used.

Phosphorus toxicity is unknown in the natural environment and phytotoxicity is not a subject of concern in disturbed soils. While forms of this element are not leached from the soil, they are complexed and low soil concentrations from an agriculturalist's point of view are to be anticipated on degraded soils in most mining regions.

Potassium

This element is essential to both plants and animals. The exact nature of the mechanism by which potassium (K) functions within plants is not known but it is considered essential for almost all plant physiologic activities.[16] Potassium plant tissue levels range from 1 to 4%.[17] This element reveals age and physiologic activity-controlled concentrations.[18] It is not uncommon to find very high potassium levels in young, vigorously growing plants and apparently deficient concentrations in senescent plant tissues of the same species or even of the same plant.

Analyses indicate that most soils have total potassium concentrations from less than 1% to 5% or more.[19,20] In disturbed soils exchangeable potassium levels are commonly determined in ammonium acetate extracts.[21,22] These concentrations usually range from 150 to 500 μg/kg dry soil. Deficient soil levels occur in eastern soils but are almost unknown in younger western soils. Analysis of any soil for toxic concentrations of potassium is not recommended.

Calcium

This element is necessary for plant and animal development. In plants it is associated primarily with cell wall structure, but it is also involved in many cellular physiologic activities. This element is immobile in plant tissues. Other than a gradual increase in concentration with the age of a tissue, calcium (Ca) levels do not show any abrupt changes related to cellular activity. The pattern of decreasing nutrient concentration as a function of a decreasing level of physiologic activity — as seen in nitrogen, phosphorus, and potassium — does not occur with calcium.

Normal plant tissue concentrations range from approximately 0.2 to 0.5% in grasses to more than 3% in many broadleaf plants.[23,24] Of equal importance to the tissue level of this element is the ratio of calcium to several other elements. These include the nonessential metals aluminum (Al) and lead (Pb) as well as the essential elements magnesium (Mg), potassium, and boron (B). Calcium ameliorates the toxic effects of high boron levels and elevated concentrations of metals such as aluminum, lead, and copper.

Total calcium ranges from less than 0.1 to 5% in most western soils, although some select soils have revealed much higher concentrations.[25,26] Calcium is the most common element on the exchange of the vast majority of soils.[27] Exceptions are (1) acid soils in which hydrogen, aluminum, or manganese (Mn) ions may dominate the exchange; (2) sodic soils in which sodium ions are very common; and

(3) serpentine soils in which the exchange may be dominated by magnesium and/or nickel. In these extreme cases in which other ions dominate the exchange they interfere with plant uptake of calcium and thereby contribute to a deficiency of this element.

In semiarid to arid soils, soluble calcium concentrations are usually determined in the aqueous extract of a soil paste.[28] Concentrations are reported to range from less than 1 to more than 30 meq/l.[29,30] Productive western sites yield values within this range. There is no concern for calcium soil deficiency or excess in disturbed soils in the West except when the pH extremes previously noted are found. When acid soils are identified, any corrective action such as liming also helps alleviate the calcium deficiency. Calcium analyses should, therefore, be carried out in semiarid regions only to determine the SAR of the soil.

Magnesium

Magnesium is a plant and animal nutrient. Although one atom of this element is found in each chlorophyll molecule, larger quantities are involved in a number of plant cellular processes.[31] Concentrations range from 0.1 to 0.5% in grasses and up to 1% in many broadleafed plant tissues. Species show strong differences in elemental levels and seasonal or age variations in tissue concentrations. These are reflections of the mobility of this element. The main concerns about magnesium levels are deficiencies, not excesses. Levels in plant tissues above 0.2% indicate an absence of deficiency symptoms.[32]

Total magnesium levels in most western soils range from a few tenths to 1 or 2%.[33,34] Soluble levels in disturbed western soils, which are usually determined in the aqueous extract of the soil paste,[35] range from less than 1 to over 20 meq/l.[36,37] Concentrations of soluble magnesium are almost always lower than calcium levels in the same soil. Apparently the soil ratio of calcium to magnesium (Ca:Mg) has little impact on plant production until soluble or exchangeable soil magnesium levels are strongly elevated.[38] When soluble soil magnesium levels are of sufficient magnitude, plant growth is inhibited. While the minimum acceptable ratio of soluble Ca:Mg was reported by one group of researchers as 1:4,[39] a second research team reported 1:5 as the point at which sensitive species began to show reduced yields.[40] Increased calcium concentrations — up to a ratio of 32:1 (Ca:Mg) — did not increase plant production more than lower elemental ratios.[41]

Serpentine soils of this country are characterized by magnesium concentrations much higher than calcium levels and often by elevated soluble nickel (Ni) levels. These soils are so toxic that they have evolved a flora unique to serpentine sites.[42,43] While these species grow slowly, they can survive in a very hostile soil environment.

Magnesium deficiency in vegetation contributes to a disease state in cattle referred to as grass tetany.[44] While not common in the Great Plains, it has occurred in Idaho, Oregon, and California. It has been suggested that the regional distribution of the disease indicates that it is closely associated with grass magnesium concentrations below 0.2%.[45]

Because magnesium soil deficiencies or toxicities are so rare in the West, analysis for this element in mine soils, overburden, or any type of soil disturbance would not be justifiable except for the determination of the SAR. Like calcium, magnesium is probably deficient in acid minesoils, but action to correct the acid problem also corrects the deficiency of magnesium unless very pure forms of lime are used.

Sulfur

Sulfur is important in plants and animals mainly as a constituent of amino acids and their resulting proteins. Plant tissue levels vary from species to species and throughout the growing season. Those plants that reveal high protein levels also have high sulfur levels. Plant tissue sulfur concentrations range from less than 0.1% to more than 1% of the plant tissue dry weight.

Total sulfur levels in soils are extremely variable but most field soils range from 100 to 500 μg/g (0.05%).[46] Certain soils, especially organic soils, may contain up to 0.5% total sulfur. No single analytical procedure to determine plant-available soil sulfur has been entirely satisfactory. Part of the problem resides in the multiple forms that sulfur may assume in a soil. Sulfur may exist as the sulfate (SO_4^{2-}), sulfite (SO_3^{2-}), or sulfide (S^{2-}). Furthermore, a major portion of the soil sulfur reserve is held in the soil organic fraction and is released slowly as this reservoir decomposes. Thus, an accurate assessment of available soil sulfur is difficult to determine and subject to question, much like plant-available soil nitrogen and phosphorus.

Toxic sulfur levels occur in plant tissues but they are not a result of elevated soil concentrations. Sulfur toxicity in plants, as indicated by elevated sulfur concentrations in tissues or by injury to plant leaves, is a result of air pollution. When plants are fumigated with sulfur dioxide (SO_2) the gas diffuses into the leaf where it is oxidized to sulfate and produces sulfuric acid. While caused by excess sulfur, the resulting injury is never due to elevated soil sulfur.

Plant deficiencies may occur in sandy, leached soils, in soils very low in organic matter, or in heavily cropped soils. However, deficiency of this element is not common in the Great Plains or most semiarid regions. Many of our western soils contain gypsum ($CaSO_4 \cdot 2H_2O$) or other oxidized sulfur compounds. Although barely soluble, they usually provide an adequate reserve for plant nutritional needs. On the other hand, sulfur deficiencies are relatively common in soils of degraded sites in more humid areas of the eastern part of the continent.

Many eastern mine spoils and some western spoils reveal large quantities of iron pyrite. In the presence of oxygen and water this mineral may oxidize and produce sulfuric acid. This contributes to the most difficult reclamation problems today: acid minesoils and acid mine drainage. The presence of reduced sulfur in surface soils, overburden, ore bodies, or subsoils underlying any disturbance is a warning that a potential acid problem may be created by manipulation of the materials. At such disturbances, further tests (e.g., buffering capacity and sulfur speciation) are warranted.

Some western states are following the example of the eastern coal producing regions. They are beginning to recognize the necessity of analyses to determine the acid-generating potential of pyritic materials. Wyoming, for example, requires acid-base analyses of overburden.[47] Such tests are extremely important to prevent the development of acid minesoils on the post-mine landscape. However, soils in the coal fields of eastern Montana and the prairie provinces of Canada contain sufficient alkalinity to buffer against acid production and pyrite is uncommon in this region. Therefore, determination of acid-base accounts in these materials is unwarranted unless problems develop.

Acid production in many active and abandoned metal mines of the West is widespread. Mine wastes and overburden at all hardrock mines in the West should be analyzed for acid-base balance. When analyses indicate a deficiency of >5 mt/ha (approximately 5 ton/acre) of base, lime incorporation is necessary for successful plant performance. More information on this topic is provided in the section on *Amendments* in Chapter 3.

TRACE ELEMENTS

The literature is inconclusive about what levels of soil trace elements cause deficiency or toxicity problems in the soil-plant environment. Concentrations may vary from one site to the next by more that ten times and neither site may express toxicity or deficiency. The data in Table 10 are taken from numerous references and different research studies.These numbers show the wide variation inherent in trace element levels in natural soils caused by different parent materials or soil genesis. Several reviews provide a thorough discussion of soil trace element levels and their relationship to parent materials or origin.[51,52]

As was pointed out earlier in this chapter, soil elemental levels can be determined by a number of methods. Strong acid or total concentrations are the most reproducible but least meaningful techniques.

TABLE 10 Trace Element Ranges Common in Soils[48–50]

Element	Ranges		Extractable[a]	
	Total	Extractable or Soluble	Low or Deficient	High or Toxic
Al	0.5–10%	Varies with pH	NA[b]	1–5
B	20–70	<1–8	0.5	8–10
Cd	0.1–2	0.01–0.05	NA[b]	Unknown
Cu	2–90	0.1–0.7	0.2	Unknown
Fe	0.3–6.8%	1.5–200	2.0	Unknown, probably never recorded in natural environment
Mn	65–1300	1–40	1.0	Unknown
Mo	0.5–20	0.05–0.3	Varies with Cu conc.	Varies with Cu conc.
Pb	3–25	0.1–3	NA[b]	Unknown
Se	2–20	0.01–0.10	Vegetation levels <0.05	Vegetation levels >0.5
Zn	10–300	0.1–14	1.0	40–80

Note: Ranges μg/g unless otherwise indicated.

[a] Fe, Mn, Cu, Zn by DTPA extraction. Mo extraction by Tamm's reagent. Al, B, Cd, Pb, Se extraction by water or hot water.

[b] Not applicable — element is not required for plant growth.

Insoluble forms of trace elements do not influence plant uptake or animal tissue levels. Of greater importance to the reclamation scientist are plant-available, extractable, or soluble levels of the various elements. The most common multiple-element extracting agents used on disturbed lands are ammonium acetate, DTPA, and AB-DTPA. Ammonium acetate extraction was designed for the slightly acid agricultural soils of eastern North America. It is adequate for the eastern and midwestern coal fields but is less reliable in semiarid to arid alkaline soils. Ammonium acetate has been replaced in the West by DTPA[53] or by AB-DTPA.[54] AB-DTPA is a modification of the DTPA extracting solution that permits the simultaneous extraction of metals (e.g., Fe, Cu, Zn, and Mg) and several anions (e.g., PO_4^{2-}, NO_3^-, and SeO_4^{2-}). Water-soluble elemental concentrations are often used in studies of toxic concentrations of metals, but reports of studies of phytotoxic levels of trace elements using any soil pool other than totals are rare.

Aluminum

There is no evidence that aluminum is required by any plant or animal even though it is one of the most common elements in the earth's crust. Plants reveal a wide range of sensitivity to this element from the very sensitive leaf lettuce *(Latuca sativa)* to the relatively tolerant bentgrass species (*Agrostis* spp.).[55] Normal, unpolluted plant tissue levels range from a few to hundreds of micrograms of aluminum per gram dry plant tissue.[56,57] Phytotoxic effects are noted in the roots before any change is observed in above-ground plant parts. Short, stunted roots are characteristic of aluminum phytotoxicity, but also result from low soil pH and elevated concentrations of many other heavy metals.

Total aluminum in soils in this country have been reported to range from 1 to 10%.[58,59] Soil solution, saturation extract, or available aluminum concentrations are closely related to the hydrogen ion concentration in the soil. At a soil pH of approximately 5.5, sufficient aluminum may be present in the soil solution (from 1 to 5 μg/g) to inhibit the growth of aluminum-sensitive plants (Figure 2). This only occurs if the soil contains enough total aluminum to produce this effect. The solubility of this element increases at pH values below 5.5, but decreases sharply as the pH increases above this value.

Natural aluminum toxicity occurs in many acid soils. It is often difficult to determine if hydrogen or aluminum ions are toxic in low pH environments or if inhibition of plant growth is simply due to low base saturation. In low pH soils, excess aluminum and hydrogen ions replace calcium and other divalent cations on the exchange sites. Because calcium is very important to plant growth, the loss of these

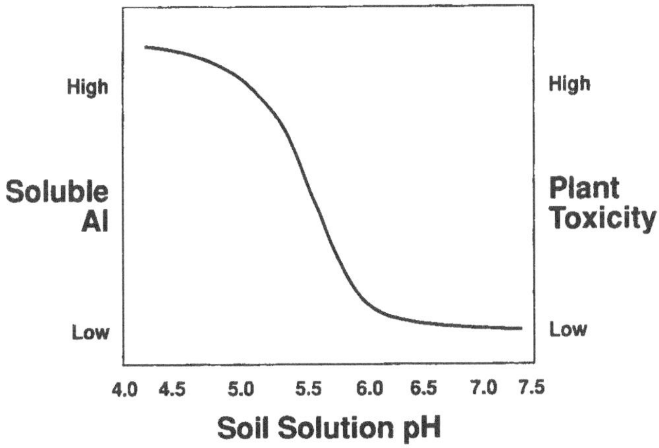

Figure 2 Generalized water-soluble aluminum toxicity curve.

ions from the exchange with their subsequent loss from the soil solution is as inhibitory to plant performance as the excess of aluminum and hydrogen ions.

Aluminum is only slightly soluble at pH values above 5.5 and below 8.5.[60] Soil concentrations of soluble forms of this element cannot occur in disturbed soils when soil pH values are within this range. Aluminum does not, therefore, pose a threat to plant growth under normal soil conditions. Analyses for this element in disturbed soils is not warranted unless the soil pH is beyond the range noted above.

Arsenic

This element, like most of the trace elements discussed in this text, is ubiquitous in the environment. It is found in all soils, plants, and animals but at very low concentrations. There is no evidence that arsenic (As) is essential to plants or animals, but several investigators[61,62] have documented increases in plant growth at very low concentrations in soils. In every study, however, higher soil levels resulted in decreased plant production. Most compounds of arsenic are toxic to plants and/or animals and many have been used as herbicides or pesticides in the past. Arsenic was previously so widely used and at such high frequencies that many old orchard soils are severely contaminated with residues of the element.[63,64] Due to decreasing use of arsenic pesticides, this element is encountered less often in the environment today than it was 50 years ago.

Arsenic is more toxic to animals than to plants. Toxicity varies, however, among animal species, valence, purity, and the chemical form of the element. Ingestion of 15 to 45 g of arsenic trioxide (As_2O_3), or only 1 to 4 g of the highly soluble sodium arsenate (Na_3AsO_4), is usually fatal to cattle.[65] The more common pentavalent forms (arsenates) are usually much less toxic than the trivalent arsenic compounds (arsenites).[66]

Normal plant tissue levels of arsenic range from almost none to 2 or 3 µg/g. Most values are less than 1 µg As per gram of dry plant tissue.[67,68] Although plant tolerance of this element varies with different species, an upper critical level for plant tissue arsenic has been listed as 20 µg/g.[69] This may be high because evidence suggests that phytotoxicity begins to be expressed at about 5 µg/g for sensitive plant species.[70]

Scientists with the U.S. Geological Survey report total soil arsenic to range from 1.2 to 97 µg/g in 490 soil samples from the western U.S.[71] Most of these values are in the lower part of this range. Soils of Russia are reported to contain from 1 to 10 µg/g dry soil.[72] Although 10 µg/g has been referenced as a maximum level for noncontaminated soils,[73] total concentrations many times this amount usually do not cause plant growth problems.

Extractable or soluble soil arsenic is a better indicator of plant response than total soil concentrations, but the most reliable extraction technique has not been clearly defined. Johnson and Barnard investigated the arsenic complexing power of ammonium acetate, salts, several acids, and numerous bases without showing a strong relationship between the concentration of arsenic in any of the extracting solutions and plant response.[74] Because of the similarity between arsenic and phosphorus in soil systems, the extraction techniques used for soil phosphorus have been suggested for arsenic. Woolson and his associates[75] evaluated six methods of soil phosphorus extraction as arsenic extractants. They recommended two techniques for the extraction of soil arsenic — one using sodium bicarbonate, and the other a combination of dilute hydrochloric and sulfuric acids.[75] These extractants were easy to use and produced good correlations between the amount of arsenic extracted and plant response, but later work by the same authors[76] indicated that different soils gave variable results. The authors attributed this variation to the influence of different soil physical and chemical properties on arsenic extractability.

Neither available soil arsenic levels in nonpolluted soils nor arsenic concentrations that are inhibitory to plant growth have been clearly established. Most studies emphasize soil additions to toxic levels and discuss total soil arsenic levels only. Such research rarely identifies extractable or available background soil levels. An early study found that water-soluble arsenic in 50 agricultural soils ranged from 0.2 to 8.9 µg/g. However, many of these soils had been treated with arsenic insecticides.[77] Presumably, a water-soluble or extractable (by Bray or other phosphate extracting method) soil arsenic concentration of less than 1 μg/g dry soil is to be expected in most natural, nonpolluted environments.

Soils that are phytotoxic because of naturally occurring arsenic have been found, but only very rarely. The probability of elevated arsenic concentrations in soils of the Great Plains or of the eastern coal fields is very remote so arsenic analyses as a routine procedure for minesoil or overburden characterization should not be undertaken.

Reports of animal poisoning by this element are always associated with human disturbances or air pollution. These poisoning events occur near primary or secondary metal processing facilities or in fields formerly treated with arsenic insecticides. Many metallic ores are arsenic containing, so vegetation or soils of disturbed or natural sites polluted by metal concentrators or smelters may contain elevated arsenic concentrations. In these areas, and in old orchard soils, elemental analysis for arsenic should be carried out to determine that levels are not elevated. Water draining from such areas should also be analyzed for arsenic during periods of peak runoff and sediment load. Hazardous waste sites associated with mining and metal processing facilities are replete with examples of elevated arsenic levels in soils and water.

Boron

Boron plays a very significant role in plant tissue development, but no realistic plant concentrations have indicated that it is toxic to animals. The role of this element in plant metabolism is unclear, but it is probably related to sugar translocation or utilization. It appears to be required by dividing and enlarging cells (e.g., buds of plants) since deficiency symptoms are observed first at these actively growing sites. The difference between the concentration of boron essential for normal plant physiologic activity and that eliciting deficiency or toxicity symptoms is small (four or five times), but the range varies from species to species. For example, 25 μg B per gram of dry ponderosa pine *(Pinus ponderosa)* needles represents a healthy background level, but 100 μg/g is clearly indicative of an injury level in this species.[78] In fourwing saltbush *(Atriplex canescens)*, tissue concentrations of several hundred micrograms per gram were found in plants not expressing any obvious toxicity symptom. Boron is immobile once deposited in plant tissues; therefore, injury caused by elevated levels occurs in older tissues where excess boron accumulates and requirements are reduced.

Total boron concentrations in most western soils range from less than 20 to 70 μg/g soil. Occasional soils reveal higher levels (300 to 500 μg B per gram) but they are not common.[79] Analyses of total soil boron concentrations are rarely required by regulatory agencies. Soil boron levels are determined in a hot water extract or in the aqueous extract of a saturation paste. Both extractions have been used in studies of toxicities or deficiencies but neither is completely satisfactory. The best correlation between plant response and soil boron concentrations are found when using soil solution boron levels. This value is difficult to determine and, therefore, the two extractions are substituted for this method. Extractable elemental levels above 5 mg/l in the saturation paste may be interpreted as a restriction on the number of plant species that can grow and reproduce in the soil. These levels are based partially upon Eaton's classic study in sand cultures.[80] While most boron-tolerant species he investigated showed injury at 10 mg/l in the nutrient solution poured over the sand, some revealed injury at 5 mg/l. Boron sensitive or semitolerant species usually showed injury between 1 and 5 mg/l. In a review by Gupta et al., the nutrient solution with which Eaton irrigated his plants was equated with a soil solution in a sandy soil.[81] These two papers form the basis for the 5 mg/l minesoil boron suspect level so common in western mine regulations.

Research by the U.S. Geological Survey[82] has shown a very strong relationship between electrical conductivity and water-soluble boron. The widespread use of this basic soil parameter (EC) as a screening tool for potential boron problems in disturbed western soils appears to be highly advisable. Multiple boron analyses of nonsaline soils at disturbed sites wastes resources and money.

Many semiarid western soils contain higher boron levels than indicated by soil water analyses. High calcium concentrations somewhat reduce the availability, and hence the toxicity, of elevated soil boron levels.[83] Boron is also adsorbed more strongly on soil particles at high pH values. Thus, liming soils or raising soil pH reduces boron availability. Acid soils with low boron concentrations may become deficient in this element after liming.

Cadmium

This element is considered nonessential for plants and animals and has been shown to be detrimental to the performance of all organisms. A notable exception is research that revealed enhanced algal growth at very low cadmium (Cd) levels in solution cultures (<0.1 μg/g).[84] Nevertheless, cadmium should be considered a toxicant and not a plant nutrient.

Background total soil cadmium levels range up to a few micrograms per gram, probably over five, but concentrations are usually less than 3 μg/g.[85,86] In western states, extractable cadmium levels are determined in AB-DTPA or DTPA soil extracts. Use of either extract has received little study but it has been reported that DTPA removed less than 0.2 μg Cd per gram of dry soil from over 100 soil samples from southeastern Montana.[87] Nevertheless, the use of DTPA- or AB-DTPA-extractable cadmium levels to establish suspect or toxic soil levels of this element is not supported by any research, so these extractants should not be used for this purpose.

Cadmium plant concentrations under nonpolluted, natural conditions are almost always less than 0.1 μg/g,[88,89] although there is much species variation. In metalliferous areas, in acid soils, or in urban centers polluted by airborne particulate fallout, cadmium concentrations in vegetation may increase to several tenths of a microgram per gram.[90,91] While mild washing removes little cadmium from plant leaves, it is not clear that plant tissue cadmium in polluted areas is entirely internal. Much of this plant burden appears to be found in particulates strongly attached to plant leaf surfaces.

The primary concern for cadmium levels in plant or animal tissues rests upon the accumulation of this element in human food chains and its impact upon man. While the amount of ingested cadmium absorbed by the human intestine is low — on the order of 5% — the half-life of the metal, once absorbed, is 20 years or more.[92] Humans, therefore, may experience cadmium intoxication at relatively low intake levels. The amount of this element permitted in the human food chain must be closely monitored and regulated. Published soil values and plant uptake rates indicate that humans are less at risk from cadmium in minesoils or other disturbed soils than from cadmium in urban air and soils. Therefore, analysis of this metal in western minesoils and overburden is not warranted by potential food chain accumulation.

Chlorine

This element is considered essential for most seed plants and higher animals but its functions in plants are not clear.[93] Chlorine is highly soluble and, as sodium chloride (NaCl), constitutes the most common component of soil salinity problems. Its solubility contributes to rapid leaching from soils and even from dead leaves and stems.

Plant tissue chlorine levels vary widely by species, concentration of salt in the root environment, and age of tissue. Leaf levels range from 0.1 to >1%, with higher accumulations in older tissues.[94–96] Halophytes such as saltgrass *(Distichlis spicata)* and greasewood *(Sarcobatus vermiculatus)* exhibit elevated salt levels (a few percent chlorine in leaves) with no observable detrimental effect. Other plants may be very sensitive to an increase of only a few micrograms per gram of chlorine in their tissues. Sensitivity in these species is expressed by marginal leaf burning, a common response to elevated plant chlorine concentrations.[97]

Total soil chlorine levels are reported to range from 60 to 200 μg/g of soil.[98] Chlorine exists as the very soluble chloride ion in the soil solution. When this elemental concentration is elevated, the impact on vegetation is usually that of total salt, not chloride ions alone. Water uptake is inhibited in response to elevated soil salt levels. Analysis of total salt is based upon the EC of the aqueous extract of the soil paste. If necessary or desirable, soluble chlorine determination may be performed on this soil water extract.[99]

The analysis of chlorine in disturbed soils is not appropriate, but the determination of the EC is necessary in western soils. Table 2 (in Chapter 1) shows the common agricultural interpretation of soil salinity levels. Western rangelands differ from common agricultural lands. Salt concentrations are usually higher in western soils than those recorded in more mesic agricultural areas. Specialists preparing revegetation programs on arid to semiarid rangelands must take this characteristic of the natural soils of the disturbed site into consideration when planning seed mixes for such soils.

Copper

This element is a plant and animal nutrient. It is a constituent of many enzymes and plays a direct role in plant and animal metabolism. Almost all plant metalloenzymes that contain copper are concerned with oxidation-reduction reactions.[100,101]

Total soil copper levels have been reported by the U.S. Geological Survey to range from 3 to 50 μg/g.[102] Other authors report similar levels. For example, a study in Colorado revealed a range of 2 to 92 μg/g in samples from several horizons in 37 different soils.[103] In this same report, almost 200 DTPA copper extractions ranged from 0.1 to 3.7 μg/g soil. A study of over 100 rangeland surface soils in southeastern Montana revealed 0.2 to 3.3 μg of DTPA-extractable Cu per gram of soil.[104] Both of these studies showed little or no change in copper concentration with increasing soil depth.

Extractable soil copper concentrations are slightly low in many western soils, probably because of the alkaline nature of these materials. Toxic copper levels do not occur in this region unless the pH of the soil is acid. There is little support for regulatory agency concern about toxic soil copper levels in minesoils of the western plains. Furthermore, toxic concentrations of AB-DTPA- or DTPA-extractable copper have not been investigated and suspect levels based upon these analytical tools should not be proposed.

Normal copper levels in vegetation range from 3 to 30 μg/g dry plant tissue.[105,106] Grasses usually reveal lower concentrations than broadleaf plants. This element is immobile in plants, yet young vigorous plant organs reveal slightly higher concentrations than older tissues. This may be due to a dilution of copper as plant tissues complete enlargement.

Concentrations of copper in forages of large areas of the northern Great Plains are borderline deficient for livestock. A normal intake of 5 to 10 μg Cu/g dry diet is recommended for cattle by the Agricultural Research Council in the U.K.[107] In the U.S., the Committee on Animal Nutrition recommended a minimum level of 4 μg/g in the dry diet for beef cattle.[108] Normal concentrations of copper in range grasses on the eastern plains of Montana are 3 to 5 μg/g.[109]

The adequacy, deficiency, or toxicity of copper in animal diets cannot be studied without a review of the molybdenum status of the animals. The two elements are closely related in livestock diets. Copper:molybdenum ratios in vegetation growing on Great Plains minesoils have been the subject of concern since publication of a report by the U.S. Geological Survey in 1978.[110] This article revealed that in white sweetclover *(Melilotus alba)* growing on mine spoils the ratio of copper to molybdenum was not adequate for ruminant diets. Subsequent studies revealed that this indeed was true. Since 1977, however, all coal mine spoils have been topsoiled and this legume is never seeded or is only a minor component of well-vegetated minesoils on the Great Plains.[111] The problem, therefore, is not of major concern on disturbed soils in the semiarid West.

Regulatory requirements for analysis of either of these elements in overburden and/or soils are hard to justify at any disturbance that receives a topsoil dressing. Imbalances between the two metals are rare, they almost always occur in alkaline soils with high watertables. It is very unlikely that these two factors occur on a disturbed site in large enough areas to pose a threat to grazing livestock.

Iron

Iron is an essential element for plants and animals. It has multiple roles in plant tissues. In its more common functions, it is an active component of many enzymes, essential for chlorophyll synthesis, and involved in electron transfer. Normal plant concentrations range from 40 μg/g (possibly lower) in many grasses, to several hundred micrograms per gram in specific forbs or shrubs.[112] For example, leaves of healthy skunkbush sumac *(Rhus trilobata)* have iron concentrations as high as 425 μg/g.[113] Concentrations in a species may also be quite variable. Multiple samples of needle-and-thread grass *(Stipa comata)* collected in early September from the northern Great Plains revealed 41, 60, 67, 80, and 142 μg Fe per gram of dry plant tissue. Each sample represented a different site and was a composite of six or more plants.

Other plants collected at the same sites indicated that a relationship existed among plant species' iron concentrations. That is, high iron levels in one plant species indicated that other species at the same site would also show high iron concentrations.

Iron is not a mobile element in plant tissues. It is similar to calcium in that tissue concentrations increase with age. There is no decrease in iron level with reduced plant physiologic activity.

Total soil levels of this element generally range from 0.5 to 10%.[114,115] Extractable levels (DTPA) are much lower. They have been reported to range from 1.5 to 160 μg/g, with a level of 4.5 μg/g identified as the minimum concentration adequate for healthy crop growth.[116] In analyses of over 100 near-surface soil samples, another study revealed only 7 DTPA iron extracts below this deficiency level.[117]

Soil iron levels that may be toxic to plants have not been studied with the DTPA extraction technique. Analysis for suspect toxic levels in soils or overburden of the Great Plains is not recommended for two reasons: (1) reclamation scientists do not know what soil levels are toxic in terms of DTPA iron extraction, and (2) the solubility of this metal is low unless the soil is acid.

Lead

Unlike cadmium and nickel, there are no reports of plant or animal growth enhancement with any levels of lead. Examples of phytotoxicity are also rare even when soil lead levels are high. On the other hand, this element is toxic to animals at relatively low concentrations. Poor plant uptake and translocation probably account for the absence of natural lead toxicities in plants and the rarity of herbivore poisoning by vegetation except in severely polluted zones. Reports of lead toxicity to animals eating contaminated foliage are relatively common around lead processing facilities, but almost all of them may be directly related to the consumption of lead-contaminated soil, particulate lead on leaf surfaces, or lead in drinking water.

Plant lead levels usually range from one to several micrograms per gram dry plant tissue in nonpolluted areas.[118,119] Soils near heavy traffic, in urban areas, land masses near metal workings, or metalliferous soils produce above-ground plant tissues with markedly elevated lead levels. These may reach several hundred micrograms of lead per gram of plant tissue with no apparent harm to some plants.[120,121] However, an Environmental Protection Agency (EPA) report suggested that the maximum lead tolerable level is 25 μg/g.[122] This was based upon an investigation that found a 10% yield reduction in barley when the tissues had Pb levels of 35 μg/g dry weight.[123]

Total soil lead levels are usually less than 100 μg/g and always less than a few hundred micrograms per gram.[124] Soil lead levels greater than these are an indication of pollution, soil contamination, or a metalliferous area.

Extractable lead concentrations (DTPA) were found to range from 0.6 to 4.0 μg/g in soils collected in southeastern Montana.[125] However, the extractability of lead from a soil is strongly pH dependent. The hydrogen ion concentration in the buffered extracting solution, therefore, has a strong influence on the amount of lead extracted by the solution. Diethylenetriaminepentacetic acid (DTPA) (buffered at pH 7.3) is the common extractant in semiarid western coal fields. Lead concentrations extracted by this solution are lower than levels extracted in the same soil by ammonium acetate (buffered at pH 7.0) or dilute hydrochloric acid. The amount of lead removed from a soil by extractants is low in western soils. It is higher in more acid eastern or forested soils. The AB-DTPA or ammonium acetate extractions of this metal have received almost no investigation at tolerable or toxic levels. Therefore, the analysis of this metal in minesoils or overburden by any extraction is of little or no value for rehabilitation.

Manganese

This element is a plant and an animal nutrient. It has a varied and poorly understood catyalytic role in plants which includes some relationship with water-splitting in photosystem II.[126] It also functions as an enzyme activator or catalyst in cells. Plant tissue concentrations are variable, but are reported to range from the low teens to slightly over 100 μg/g in most grasses, forbs, and shrubs.[127] However, a report from the northern Great Plains indicated that 17 samples of smooth brome *(Bromus inermis)* averaged 144 μg Mn per gram of dry plant tissue and 14 samples of willow leaves (*Salix* sp.) averaged 633 μg/g.[128] The U.S. Geological Survey reported that manganese concentrations of 400 to 500 μg/g usually indicates a plant toxicity problem.[129]

Soil manganese concentrations are as variable as plant levels. The vast majority of total surficial soil manganese concentrations range from 100 to over 1000 μg/g.[130] Studies of strong-acid-extractable soil manganese in the coal-producing areas of the northern Great Plains revealed concentrations ranging from 150 to almost 500 μg/g.[131]

Extractable manganese concentrations may be determined with a number of extracting agents. These include ammonium acetate, AB-DTPA, and DTPA. Levels of manganese extracted by DTPA ranged from 0.6 to 39.5 μg/g in 37 Colorado soils.[132] Surface soils from undisturbed Montana grasslands revealed a range of 6 to 47 μg DTPA-extractable Mn per gram of soil.[133] Both reports showed a decrease of extractable manganese with increasing depth in the soil profile.

Plant deficiencies and toxicities occur in natural soils, but they are almost always related to pH or state of oxidation. In general, soil manganese levels are toxic only if the soil pH is below 5.5 or if the soils are strongly reduced. A concentration of 2.5 μg water-soluble Mn per gram of acid soil (pH <5.5) was reported to be toxic to plants.[134] Ammonium acetate-extractable manganese was always toxic to alfalfa when the elemental concentration exceeded 50 μg/g of soil.[135] Toxicity levels generated from DTPA-extractable manganese have not been researched and, therefore, are unknown. The soil level at which Colorado agricultural crops would show a growth response was reported as less than 1.0 μg DTPA-extractable Mn per gram of soil.[136]

Manganese toxicities are often reported in acid minesoils but these may be corrected by lime application. Residual acid-generating capacity may reacidify a limed soil and release soluble manganese again. It is not clear that manganese toxicity is always an excess of the element. The presence or absence of adequate iron concentrations is also important. An iron to manganese ratio higher than 1.5 usually indicates a healthy soil/plant system. Lower ratios are cause for concern.[137]

Mercury

This element is both acutely and chronically toxic to animals in any of its three forms: elemental mercury (Hg), inorganic salts, and organic mercury compounds. Methylmercury may produce fatalities in humans or animals if ingested at rates measured in milligrams. The most highly toxic mercury compounds (methyl- and ethylmercury) are produced by bacterial action on the other compounds of mercury. Thus, any form of mercury must be viewed with concern.

D'Itri reported that total soil mercury levels were in the range 10 to 150 μg/kg.[138] The U.S. Geological Survey reported a broader range, but over 70% of their samples fell between 10 and 150 μg/kg.[139] These levels may increase into the microgram per gram range in heavily polluted environments of cities, near metal reduction works, or over ore bodies. Concentrations of this element in soils are so low that research to relate extractable metal levels and plant response has not been located.

Mercury is not a plant nutrient and, fortunately, plants do not readily absorb this element from their environment nor do they readily translocate it. Nevertheless, mercury applied as a fungicide to seed coats has been shown to be transported to above-ground plant parts and to be concentrated in various components of the food chain. The widespread use of mercury compounds as seed coatings is no longer permitted because it contaminates human and animal food chains.

Mercury levels in nonpolluted areas are usually less than 0.5 μg/g dry plant tissue. Roots and green tissues of plants found growing in soil over a cinnabar vein (mercury ore) yielded mercury concentrations of only 1 to 3.5 μg/g.[140] Because of the low background levels of this element and poor plant uptake, concerns about this metal in minesoils and overburden are unwarranted except in metalliferous or polluted areas.

Molybdenum

This element is an essential nutrient for plants and animals. In plants it is involved in nitrate reduction and nitrogen fixation. It is found in several enzymes in animals.

Normal soil molybdenum levels range from <3 to >40 μg total Mo per gram of dry soil.[141] Although numerous extraction techniques (water, ammonium carbonate, ammonium acetate, etc.) have been evaluated with respect to molybdenum, the ammonium oxalate-oxalic acid extraction[142] is most commonly referred to in disturbed land soil extraction procedures. It is the only method that has been field evaluated and for which any relationships between soil concentrations and plant response have been formulated.[143] It is the generally accepted extractant for western soils. Molybdenum levels as determined by this technique generally have been reported in the few tenth microgram per gram range.[144]

Soil concentrations low enough to inhibit plant development have been recorded but they are rare. The most consistent feature of molybdenum-deficient soils has been low pH.[145] Apparently this element is more firmly adsorbed in low pH soils although root absorption is also enhanced in this environment. Soil levels high enough to produce toxic reactions in plants have not been recorded in natural environments. Elevated soil levels are found in undisturbed environments, but they are identified by observation of animals and then soil and/or plant analyses.

Normal plant molybdenum concentrations range from a few tenths to micrograms per gram of dry plant tissue.[146,147] Higher levels are found in legumes. Of several legumes analyzed in a study in Montana, white sweetclover had the highest average molybdenum concentration (5.7 μg/g).[148] Unfortunately, plants may accumulate sufficient molybdenum to produce a toxic response in grazing animals without expressing any symptoms themselves. Molybdenum concentrations that are toxic to grazing animals are closely related to copper concentrations in the animal's diet. The two elements must be evaluated together to understand animal needs for either element.

Molybdenosis is an induced copper deficiency caused by elevated concentrations of molybdenum in the animal diet. In general, a dietary copper:molybdenum ratio that is less than 5:1 is cause for concern, while ratios less than 2:1 clearly indicate an animal maintenance problem.

It is difficult to make definitive statements about the levels of copper or molybdenum in plant tissues and animal health because other dietary and environmental factors influence the characteristic molybdenosis syndrome. For example, copper may be stored in the animal body for rather long periods and utilized when needed. In addition, copper is much less available in fresh herbage than in cured forage. The presence of elevated sulfate levels also impacts both metal levels. Soil sulfate reduces the absorption or utilization of molybdenum by plants. Plant sulfate complexes copper in the intestinal tract of an animal and thereby reduces copper absorption into the blood system.[149] This confusing situation is somewhat clarified in Table 11. In this table the author separates ruminant diets according to their copper and molybdenum contents and accompanying disease syndrome. Molybdenosis, therefore, is a disease state in grazing animals produced by an imbalance of copper and molybdenum in the diet. The severity of the disease may be enhanced or lessened by the presence of sulfate in the soil or in forages consumed by the affected animals.

Nickel

While considered solely a toxic element by many, nickel has been shown to be an essential element to at least one species of algae,[151] to chickens,[152,153] and probably to swine.[154] It is needed in such small quantities that purified diets deficient in this element are difficult to prepare. The functions of nickel in plants and animals are unknown.

All soils contain some quantity of nickel. Total soil levels generally range between 3 and 30 μg/g.[155] The only known occurrences of naturally toxic nickeliferous materials are associated with serpentine soils.[156,157] Total soil nickel concentrations in serpentine areas range from a few hundred to 9700 μg/g of soil.

Several methods of soil extraction have been utilized for this element. In most extractions the concentration of nickel is pH dependent.

TABLE 11 Livestock Response to Cu and Mo in Their Diet[150]

	Elemental concentrations in diet (μg/g dry wt)	
State	Cu	Mo
Cu toxicity	>50	Almost any level
Normal	5–15	1–5
Molybdenosis	5–8	>5
Cu deficiency	<5	Almost any level

Diethylenetriaminepentacetic acid is commonly utilized for nickel extraction in western soils. Extractable nickel levels (DTPA) were reported to range from 0.2 to 1 μg/g in surface soils from southeastern Montana.[158]

As with most other metals, plants show species, tissue, and seasonal variation in nickel levels. Nevertheless, elemental concentrations in plants rarely exceed 2 to 3 μg/g.[159,160] Nickel levels above 5 μg/g dry plant tissue indicate acid or serpentine soils or pollution. For most species and populations, nickel concentrations between 20 and 30 μg/g usually inhibit plant growth.[161] However, researchers have reported 30 to 230 μg/g plant tissue in vegetation growing on serpentine asbestos mine wastes in Canada.[162] This vegetation was showing the effects of numerous toxicities, so it was not possible to isolate how nickel affected the plants.

The ratio of Ni:Fe in plant tissues has been suggested as a better indicator of the toxic effects of nickel than the concentration of the metal alone.[163] A Ni:Fe ratio of 1.0 has been reported to be an indicator of toxic impacts of this element on plants.

Because of the pH dependence of nickel solubility, levels in alkaline western soils are usually low. References to phytotoxicity studies utilizing any availability index for nickel are rare, and toxic levels of DTPA-extractable nickel have not been defined. The use of DTPA to determine toxic or suspect soil levels for this element is not supported by any research and, therefore, should not be incorporated into minesoil analytical programs.

Selenium

This element is an essential animal nutrient but evidence that it is necessary for the growth or development of plant species is very weak. While selenium has been shown to enhance the development of a small group of plant species that absorb large quantities of the element, it has not been shown to have any effect on the vast majority of plant species.[164] Selenium is absorbed by plants from the soil and retained in various tissues. However, the element is volatile and may be lost from vegetation during hot summer days. A characteristic seleniferous odor has been noted by scientists and others working with selenium-containing plants in the field and laboratory.

Total selenium concentrations in noncontaminated soils may range up to 3 μg/g, but are usually less than 1 μg/g of dry soil.[165] Even in seleniferous western U.S. soils, total selenium concentrations are usually less than 10 μg/g although concentrations up to 100 μg/g have been reported in the Niobrara silt loams of Wyoming.[166]

Studies of soil-plant selenium relationships with numerous extractants have had limited success.[167,168] The two most common selenium extracting agents used today are hot water and AB-DTPA. Levels determined by these two extractants are usually low (from 0.01 to 0.1 μg/g) in western soils and overburden.

The classic works on selenium as reported by Trelease and Beath[169] and by Rosenfeld and Beath[170] utilized the hot water-soluble extraction technique to determine plant-available soil selenium. This water-soluble form was identified as the selenate, or oxidized selenium. Early greenhouse studies of the response of plants to soil selenium showed a strong relationship between plant selenium concentration and water-soluble or selenate selenium.[171] Researchers are still using this technique today. Unfortunately, the preponderance of water-soluble selenium extractions made in commercial and research laboratories today indicate that disturbed western soils and overburden contain little or no available selenium. Most researchers agree that these results are surprisingly low. Selenium concentrations in the Fort Union and other formations overlying western coal beds should reveal some elevated levels. Evidently, the hot water-soluble selenium extraction technique does not lend itself to commercial modification. Interlaboratory comparisons made by two western regional study teams have shown that most commercial laboratories and many research laboratories do not get accurate or precise selenium determinations with this extraction technique.[172,173]

A possible improvement of this soluble selenium extraction procedure is the AB-DTPA soil selenium extraction. Scientists working with AB-DTPA reported a very strong correlation between hot water-soluble and AB-DTPA-extractable soil selenium.[174] They also reported a stronger relationship between AB-DTPA-extractable soil selenium and plant tissue selenium levels than between water-soluble selenium and plant tissue selenium. The AB-DTPA extraction is apparently more precise and extracts slightly more selenium than hot water. This extraction technique is receiving increasing use and experimentation in research and commercial laboratories today. Nevertheless, many regulatory agencies and laboratories continue to recommend and use the water-soluble selenium extraction technique. This is at least partially because almost all past animal and plant toxicity or deficiency studies were based on this technique.

Studies of soil concentrations that may produce seleniferous vegetation have been successfully carried out in glass houses.[175,176] These studies suggest that certain plant species accumulate sufficient selenium to be toxic to animals if soluble soil selenium (selenate) concentrations are only a few tenths of a microgram per gram.[177] However, attempts to develop relationships between plant and soil selenium levels under field conditions have not been successful. A recent field study of seleniferous soils and native vegetation[178] revealed that AB-

DTPA-extractable soil selenium concentrations of less than 0.2 μg/g produced plant tissue selenium levels greater than levels recommended for prolonged consumption by livestock. Because this study investigated plant selenium concentrations only on seleniferous soils, it should be extrapolated to nonseleniferous soils with caution.

Tissue selenium concentrations in most forage plant species typically range from 0.01 to less than 1.0 μg/g dry weight.[179] Plant levels below 0.1 μg/g are considered deficient for livestock nutritional needs (Table 12). On such a diet ruminants develop a disease state referred to as white muscle disease.[180,181] This disease is found in animals shortly after birth and is usually fatal unless corrected immediately. Plant selenium concentrations above 3 to 5 μg/g of dry plant tissue may produce toxic symptoms in livestock.[182,183] Table 12 summarizes available dietary information on selenium.

Most rangeland vegetation never reveals selenium levels high enough to produce toxicity symptoms in livestock, but certain plant species almost always contain elevated selenium concentrations. This finding lead Beath and Trelease to propose three classes of vegetation with respect to selenium accumulations.[184] These classes were

1. Selenium accumulators or indicators. These plants include species within the genera *Stanleya*, *Haplopappus*, *Xylorrhiza*, and *Astragalus*. As a general rule these plants retain sufficient selenium in their tissues throughout the growing season to be acutely toxic if consumed by livestock. Concentrations of several hundred to a few thousand micrograms selenium per gram of dry plant tissue have been reported in these plants and they are always found on seleniferous soils. The relationship between these species and seleniferous soils is so strong that Beath and his co-workers[185] used the plants to map seleniferous soils.
2. Secondary selenium accumulators or intermediate selenium-bearing plants. These genera are more numerous than the selenium indicator plants and usually reveal lower or more erratic selenium concentrations. Nevertheless, selenium levels in this group may be high enough to produce chronic toxicity problems in livestock. Several examples in this category are found in the genera *Atriplex*, *Grindelia*, *Aster*, *Machaeranthera*, *Penstemon*, and *Gutierrezia*. Selenium concentrations in this group rarely exceed 100 μg/g.
3. Nonaccumulating plants. The vast majority of semiarid rangeland plants fall into a third group. These plants do not accumulate sufficient selenium to pose a threat to livestock unless the soil concentration of the element is extremely high (several tenths of a microgram per gram). Selenium concentrations in this group rarely exceed 5 to 10 μg/g and usually range from <0.1 to 1 μg/g.

TABLE 12 Nutritional Se Levels ($\mu g/g$ Dry Diet)

Physiological Condition	Dietary Se Concentrations
Deficiency	<0.1
Adequacy	0.1–1.0
Chronic toxicity	3–20
Acute toxicity	>100

The problem of actual or potential seleniferous soils remains for the regulator and the reclamation specialist. Investigators have known for many years that the identification of potential or actual seleniferous soils is very difficult. The present program required by federal and state regulations in western coal fields is the most rigorous system proposed for the assessment of selenium in mined lands to date, but it does not identify the water-insoluble selenium forms in soils or overburden. These reduced or insoluble selenium forms — selenites, selenides, and elemental selenium — are present in at least some overburden materials. Neither of the presently used extraction techniques provides any evidence of the concentration of these reduced forms in soil or overburden. Reduced selenium compounds may be oxidized to soluble selenates in the aerated environment of the upper levels of an inverted, disturbed soil profile. While it is clear that these chemical changes take place, we do not know the rates at which these transformations occur[186] and we cannot determine when, if ever, these soils will become seleniferous.

The identification of actual or potential seleniferous minesoils remains unresolved. The hot-water-soluble selenium extraction lacks precision and accuracy even when soluble selenium concentrations exceed levels that produce seleniferous vegetation. The AB-DTPA extraction technique, while extracting more selenium than hot water, has not received sufficient review and study to permit complete reliance upon this new technique. If the AB-DTPA extraction technique is rigorous enough for routine soil selenium extractions, like the hot water extraction method, it does not extract insoluble selenium forms.

It is possible that we are unknowingly creating seleniferous rootzone materials on western minesoils or other disturbances such as highway rights-of-way. In the absence of an analytical tool to identify potential seleniferous materials, it is important that the seeding of selenium indicator or accumulator species should not be carried out on disturbed sites if there is any possibility of elevated levels of selenide, selenite, or elemental selenium in the rootzone. Future animal toxicity problems that these plants pose can be avoided if the new plant community seeded on the disturbance does not include them, and if their invasion of the site is prevented.

Sodium

This element is an essential animal nutrient which is required by some plants. The species in which sodium elicits a growth response are generally halophytes such as *Atriplex vesicaria*,[187] *A. nummularia*,[188] and *Halogeton glomeratus*.[189] The exact function of sodium in these and other plants is not clear.

Total soil sodium concentrations usually range from 0.1% to a few percent.[190] Lower concentrations are found in humid areas where soils are leached, and higher levels in semiarid to arid regions. Sodium is measured in the aqueous extract of a saturated paste although it may also be determined in extractants such as ammonium acetate or DTPA. The effect of elevated sodium on a plant is not directly upon the organism's tissues but rather indirectly, through its impact upon the soil system. Soil elemental levels that pose problems to plant germination, establishment, and growth are related to the proportion of calcium and magnesium in the soil as well as the texture of the soil material. These subjects were addressed under the heading *Sodium in Soils* in Chapter 1.

Zinc

This element is an essential nutrient for plants and animals. It is necessary for the function, structure, and/or regulation of many diverse enzymes. Among other activities in plants, zinc is required for chloroplast development, nucleic acid and protein synthesis, and auxin stability.[191] Deficiency or toxicity may result in severe metabolic disorders in plants or animals.

Total zinc contents of western soils range from 10 to 300 μg/g for most uncontaminated surface materials.[192] A range of DTPA-extractable zinc of 0.13 to 14.2 μg/g in Colorado soils was reported.[193] Investigations in Montana prairie soils showed a Zn range of 0.3 to 4.3 μg/g dry soil.[194]

Most research with zinc soil levels has concerned deficiencies rather than toxicities. The Colorado study reported a crop adequacy level around 1.0 μg DTPA-extractable zinc per gram of soil.[195] Other researchers have reported similar values.[196,197] Zinc soil toxicities are rarely found in natural soils and fewer toxicities have been investigated with DTPA as the elemental extracting solution. When found, zinc soil problems are usually a result of human activities producing acid mine drainage or polluting soils.

Total soil zinc concentrations producing phytotoxic responses are usually in the 250 to 650 μg/g range.[198] Although investigators have reported that 10 to 14 μg DTPA-extractable zinc per gram of soil reduced nitrogen fixation in pea plants,[199] most DTPA zinc extractions reveal phytotoxic responses between 50 and 125 μg/g.

Normal plant tissue levels of zinc range from 10 to 50 μg/g dry plant tissue.[200,201] Concentrations in the lower part of this range are found in grasses. Plant tissue levels, usually considered indicative of toxic conditions within the plant, vary among reports from researchers. Phytotoxic plant tissue ranges from 200 to 300 μg zinc per gram of dry weight were reported by the U.S. Geological Survey,[202] while other investigators have reported phytotoxicity at plant tissue concentrations from 300 to 400 μg/g.[203]

In view of the phytotoxic soil zinc levels reported above, and the tissue levels required to produce a toxic response in plants, the analysis for potential zinc toxicity in disturbed soils is not appropriate in semiarid to arid regions. When soils have been impacted by smelter emissions, mine wastes, or acid-producing pyritic materials, analyses of zinc and other metals are appropriate.

REFERENCES

1. Jones, U. S., *Fertilizers and Soil Fertility*, 2nd ed., Prentice-Hall, Englewood Cliffs, NJ, 1982, 3.
2. Leisman, G. A., A vegetation and soil chronosequence on the Mesabi iron range spoil banks, Minnesota, *Ecol. Monogr.*, 27, 221, 1957.
3. Subcommittee on Feed Composition and Committee on Animal Nutrition, *Atlas of Nutritional Data on United States and Canadian Feeds*, National Academy of Sciences, Washington, D.C., 1971, 5.
4. Murray, R. B., Mayland, H. F., and VanSoest, P. J., Growth and Nutritional Value to Cattle of Grasses on Cheatgrass Range of Southern Idaho, For. Serv. Res. Paper INT-199, U.S. Department of Agriculture, Washington, D.C., 1978, 22.
5. Miller, R. M. and May, S. W., Below-ground ecosystem project, in *Land Reclamation Program, Annual Report, 1978*, Argonne National Laboratory, Argonne, IL, 1979, 64.
6. Ross, D. J. and Cairns, A., Nitrogen availability and microbial biomass in stockpiled topsoils in Southland, *N.Z. J. Sci.*, 24, 137, 1981.
7. Visser, S., Fujikawa, J., Griffith, C. L., and Parkinson, D., Effect of topsoil storage on microbial activity, primary production, and decomposition potential, *Plant Soil*, 82, 41, 1984.
8. McGinnies, W. J. and Crofts, D. A., Effects of N and P fertilizer placement on establishment of seeded species on redistributed mine topsoil, *J. Range Manage.*, 39, 118, 1986.
9. Reeder, J. D. and Sabey, B., Nitrogen, in *Reclaiming Mine Soils and Overburden in the Western United States*, Williams, R. D. and Schuman, G. E., Eds., Soil and Water Conservation Society, Ankeny, IA, 1987, 155.

10. Dreher, G. B., Finkelman, P. B., and Sandvik, E. I., Factors Affecting Ground Water Quality in the Caballo Mine Backfill: Status Report, Paper prepared by Exxon Products Research Co., Houston, TX, 1985, 3.
11. Subcommittee on Feed Composition and Committee on Animal Nutrition, *Atlas of Nutritional Data on United States and Canadian Feeds*, National Academy of Sciences, Washington, D.C., 1971, 5.
12. Murray, R. B., Mayland, H. F., and VanSoest, P. J., Growth and Nutritional Value to Cattle of Grasses on Cheatgrass Range of Southern Idaho, For. Serv. Res. Paper INT-199, U.S. Department of Agriculture, Washington, D.C., 1978, 24.
13. Olsen, S. R. and Sommers, L. E., Phosphorus, in *Methods of Soil Analysis, Part 2*, 2nd ed., Page, A. L., Ed., American Society of Agronomy, and Soil Science Society of America, Madison, WI, 1982, 403.
14. Soltanpour, P. N., Follett, R. H., and Westfall, D. G., *Guide to Fertilizer Recommendations in Colorado*, Coop. Ext. Serv., Colorado State University, Fort Collins, 1985, 3.
15. Sandoval, F. M. and Power, J. F., Laboratory Methods Recommended for Chemical Analysis of Mined-Land, Spoils and Overburden in Western United States, Handbook No. 525, U.S. Department of Agriculture, Washington, D.C., 1977, 13.
16. Clarkson, D. T. and Hanson, J. B., The mineral nutrition of higher plants, *Annu. Rev. Plant Physiol.*, 31, 239, 1980.
17. Subcommittee on Feed Composition and Committee on Animal Nutrition, *Atlas of Nutritional Data on United States and Canadian Feeds*, National Academy of Sciences, Washington, D.C., 1971, 5.
18. Murray, R. B., Mayland, H. F., and VanSoest, P. J., Growth and Nutritional Value to Cattle of Grasses on Cheatgrass Range of Southern Idaho, For. Ser. Res. Paper INT-199, U.S. Department of Agriculture, Washington, D.C., 1978, 30.
19. Shacklette, H. T., Hamilton, J. C., Boerngen, J. G., and Bowes, J. M., Elemental Composition of Surficial Materials in the Conterminous United States, Prof. Paper 574-D, U.S. Geological Survey, Washington, D.C., 1971, D3.
20. Connor, J. J. and Shacklette, H. T., Background Geochemistry of Some Rocks, Soils, Plants, and Vegetables in the Conterminous United States, Prof. Paper 574-F, U.S. Geological Survey, Washington, D.C., 1975, 120.
21. Richards, L. A., *Diagnosis and Improvement of Saline and Alkali Soils*, Agric. Res. Serv. Handbook No. 60, U.S. Department of Agriculture, Washington, D.C., 1969, 19.
22. Knudsen, D., Peterson, G. A., and Pratt, P. F., Lithium, sodium, and potassium, in *Methods of Soil Analysis, Part 2*, 2nd ed., Page, A. L., Ed., American Society of Agronomy and Soil Science Society of America, Madison, WI, 1982, 225.
23. Loneragan, J. F., Gladstones, J. S., and Simmons, W. J., Mineral elements in temperate crop and pasture plants. II. Calcium, *Aust. J. Agric. Res.*, 19, 353, 1968.

24. Hamilton, J. W. and Gilbert, C. S., *Composition of Wyoming Range Plants and Soils,* Agric. Exp. Sta. Res. J. 55, University of Wyoming, Laramie, 1972, 6.
25. Shacklette, H. T., Hamilton, J. C., Boerngen, J. G., and Bowes, J. M., Elemental Composition of Surficial Materials in the Conterminous United States, Prof. Paper 574-D, U.S. Geological Survey, Washington, D.C., 1971, D3.
26. Connor, J. J. and Shacklette, H. T., Background Geochemistry of Some Rocks, Soils, Plants, and Vegetables in the Conterminous United States, Prof. Paper 574-F, U.S. Geological Survey, Washington, D.C., 1975, 44.
27. Thomas, G. W., Exchangeable cations, in *Methods of Soil Analysis,* Part 2, 2nd ed., Page, A. L., Ed., American Society of Agronomy and Soil Science Society of America, Madison, WI, 1982, 159.
28. Lanyon, L. E. and Heald, W. R., Magnesium, calcium, strontium, and barium, in *Methods of Soil Analysis, Part 2,* 2nd ed., Page, A. L., Ed., American Society of Agronomy and Soil Science Society of America, Madison, WI, 1982, 245.
29. Bradford, G. R., Bair, F. L., and Hunsaker, V., Trace and major element contents of soil saturation extracts, *Soil Sci.,* 112, 225, 1971.
30. Munshower, F. F., DePuit, E. J., and Neuman, D. R., *The Effects of Stack Emissions on the Range Resource in the Vicinity of Colstrip, Montana: Pre-Operational Description of the Range Ecosystem,* MT Agric. Exp. Sta. Res. Rep. 126, Montana State University, Bozeman, 1978, 18.
31. Clarkson, D. T. and Hanson, J. B., The mineral nutrition of higher plants, *Annu. Rev. Plant Physiol.,* 31, 239, 1980.
32. Embleton, T. W., Magnesium, in *Diagnostic Criteria for Plants and Soil,* Chapman, H. O., Ed., University of California, Riverside, 1966, 225.
33. Shacklette, H. T., Hamilton, J. C., Boerngen, J. G., and Bowes, J. M., Elemental Composition of Surficial Materials in the Conterminous United States, Prof. Paper 574-D, U.S. Geological Survey, Washington, D.C., 1971, D3.
34. Connor, J. J. and Shacklette, H. T., Background Geochemistry of Some Rocks, Soils, Plants, and Vegetables in the Conterminous United States, Prof. Paper 574-F, U.S. Geological Survey, Washington, D.C., 1975, 91.
35. Lanyon, L. E. and Heald, W. R., Magnesium, calcium, strontium, and barium, in *Methods of Soil Analysis Part 2,* 2nd ed., Page, A. L., Ed., American Society of Agronomy and Soil Science Society of America, Madison, WI, 1982, 247.
36. Bradford, G. R., Bair, F. L., and Hunsaker, V., Trace and major element contents of soil saturation extracts, *Soil Sci.,* 112, 225, 1971.
37. Munshower, F. F., DePuit, E. J., and Neuman, D. R., *The Effects of Stack Emissions on the Range Resource in the Vicinity of Colstrip, Montana: Pre-Operational Description of the Range Ecosystem,* MT Agric. Exp. Sta. Res. Rep. 126, Montana State University, Bozeman, 1978, 18.
38. Hunter, A. S., Yield and composition of alfalfa as affected by variations in the calcium-magnesium ratio in the soil, *Soil Sci.,* 67, 53, 1949.
39. Hunter, A. S., Yield and composition of alfalfa as affected by variations in the calcium-magnesium ratio in the soil, *Soil Sci.,* 67, 53, 1949.

40. Walker, R. B., Walker, H. M., and Ashworth, P. R., Calcium magnesium nutrition with special reference to serpentine soils, *Plant Physiol.*, 30, 214, 1955.
41. Hunter, A. S., Yield and composition of alfalfa as affected by variations in the calcium-magnesium ratio in the soil, *Soil Sci.*, 67, 53, 1949.
42. Kruckeberg, A. R., The ecology of serpentine soils. III. Plant species in relation to serpentine soils, *Ecology*, 35, 267, 1954.
43. Walker, R. B., Walker, H. M., and Ashworth, P. R., Calcium magnesium nutrition with special reference to serpentine soils, *Plant Physiol.*, 30, 214, 1955.
44. Kubota, J., Oberly, G. H., and Naphan, E. A., Magnesium in grasses of three selected regions in the United States and its relationship to grass tetany, *Agron. J.*, 72, 907, 1980.
45. Kubota, J., Oberly, G. H., and Naphan, E. A., Magnesium in grasses of three selected regions in the United States and its relationship to grass tetany, *Agron. J.*, 72, 907, 1980.
46. Beaton, J. D., Burns, G. R., and Platou, D. J. *Determination of Sulfur in Soils and Plant Materials*, Tech. Bull. No. 14, Sulfur Institute, Washington, D.C., 1968, 2.
47. Wyoming Department of Environmental Quality, Topsoil and overburden, Guideline No. 1, Land Qual. Div., Cheyenne, 1984, 12.
48. Follett, R. H. and Lindsay, W. L., *Profile Distribution of Zinc, Iron, Manganese, and Copper in Colorado Soils*, Exp. Sta. Tech. Bull. 110, Colorado State University, Fort Collins, 1970, 10.
49. Connor, J. J. and Shacklette, H. T., Background Geochemistry of Some Rocks, Soils, Plants, and Vegetables in the Conterminous United States, Prof. Paper 574-F, U.S. Geological Survey, Washington, D.C., 1975, 26.
50. Munshower, F. F., DePuit, E. J., and Neuman, D. R., *The Effects of Stack Emissions on the Range Resource in the Vicinity of Colstrip, Montana: Pre-Operational Description of the Range Ecosystem*, MT Agric. Exp. Sta. Res. Rep. 126, Montana State University, Bozeman, 1978, 15.
51. Kabata-Pendias, A. and Pendias, H., *Trace Elements in Soils and Plants*, CRC Press, Boca Raton, FL, 1984.
52. Bowen, H. J., *Trace Elements in Biochemistry*, Academic Press, New York, 1966.
53. Lindsay, W. L. and Norvell, W. A., A micronutrient soil test for Zn, Fe, Mn, and Cu, *Agron. Abstr.*, 84, 1969.
54. Soltanpour, P. N. and Schwab, A. P., A new soil test for simultaneous extraction of macro- and micro-nutrients in alkaline soils, *Commun. Soil Sci. Plant Anal.*, 8, 5, 1977.
55. McLean, F. T. and Gilbert, B. E., The relative aluminum tolerance of crop plants, *Soil Sci.*, 24, 163, 1927.
56. Shaw, G., Concentrations of twenty-eight elements in fruiting shrubs downwind of the smelter at Flin Flon, Manitoba, *Environ. Pollut. Ser. A*, 25, 197, 1981.
57. Gough, L. P. and Severson, R. C., Element Concentrations in Rehabilitation Species From Thirteen Coal-Stripmines in Five Western States and Alaska, Open-file Rep. 81–182, U.S. Geological Survey, Denver, CO, 1981, 24.

58. Shacklette, H. T., Hamilton, J. C., Boerngen, J. G., and Bowes, J. M., Elemental Composition of Surficial Materials in the Conterminous United States, Prof. Paper 574-D, U.S. Geological Survey, Washington, D.C., 1971, D3.
59. Connor, J. J. and Shacklette, H. T., Background Geochemistry of Some Rocks, Soils, Plants, and Vegetables in the Conterminous United States, Prof. Paper 574-F, U.S. Geological Survey, Washington, D.C., 1975, 22.
60. Bache, W. W., Soil acidification and aluminum, *Soil Use Manage.*, 1, 10, 1985.
61. Liebig, G. F., Bradford, G. R., and Vanselow, A. V., Effects of arsenic compounds on citrus plants in solution culture, *Soil Sci.*, 88, 342, 1959.
62. Jacobs, W. W., Keeney, D. R., and Walsh, L. M., Arsenic residue toxicity to vegetable crops grown on Plainfield sand, *Agron. J.*, 62, 588, 1970.
63. Overley, F. L., *Cover Crops in Apple Orchards on Arsenic-Toxic Soils*, Agric. Exp. Sta. Bull. No. 514, Washington State University, Pullman, 1950, 3.
64. Woolson, E. A., Axley, J. H., and Kearney, P. C., The chemistry and phytotoxicity of arsenic in soils. I. Contaminated field soils, *Soil Sci. Soc. Am. Proc.*, 35, 938, 1971.
65. Clarke, E. G. and Clarke, M. L., Arsenic, in *Garner's Veterinary Toxicology*, 3rd ed., Williams & Wilkins, Baltimore, 1967, 44.
66. Franke, K. W. and Moxon, A. L., A comparison of the minimum fatal dose of selenium, tellurium, arsenic, and vanadium, *J. Pharmacol. Exp. Ther.*, 58, 454, 1936.
67. Gough, L. P. and Severson, R. C., Element Concentrations in Rehabilitation Species From Thirteen Coal-Stripmines in Five Western States and Alaska, Open-file Rep. 81–182, U.S. Geological Survey, Denver, CO, 1981, 24.
68. Schroeder, H. A. and Balassa, J. J., Abnormal trace metals in man, Arsenic, *J. Chronic Dis.*, 19, 85, 1966.
69. Davies, B. E., Beckett, P. H., and Wollan, E., Critical levels of twenty potentially toxic elements in young barley, *Plant Soil*, 49, 395, 1978.
70. Environmental Protection Agency, Assessment of the Toxicity of Arsenic, Cadmium, Lead, and Zinc in Soil, Plants, and Livestock in the Helena Valley of Montana, Contract No. 68–01–7251, U.S. Environmental Protection Agency, Helena, MT, 1987, 87.
71. Shacklette, H. T., Boerngen, J. G., and Keith, J. R., Selenium, Fluorine, and Arsenic in Surficial Materials of the Conterminous United States, Circ. 692, U.S. Geological Survey, Washington, D.C., 1974, 10.
72. Swaine, D. J., *The Trace Element Content of Soils*, Commonwealth Agriculture Bureau, Farnham Royal, Bucks, U.K., 1955, 2.
73. Adriano, D. C., *Trace Elements in the Terrestrial Environment*, Springer-Verlag, New York, 1986, 51.
74. Johnson, J. E. and Barnard, W. M., Comparative effectiveness of fourteen solutions for extracting arsenic from four western New York soils, *Soil Sci. Soc. Am. Proc.*, 43, 304, 1979.
75. Woolson, E. A., Axley, J. H., and Kearney, P. C., Correlation between available soil arsenic estimated by six methods and response of corn, *Soil Sci. Soc. Am. Proc.*, 35, 101, 1971.

76. Woolson, E. A., Axley, J. H., and Kearney, P. C., The chemistry and phytotoxicity of arsenic in soils. II. Effects of time and phosphorus, *Soil Sci. Soc. Am. Proc.*, 37, 254, 1973.
77. Greaves, J. E., The arsenic content of soils, *Soil Sci.*, 38, 355, 1934.
78. Munshower, F. F. and Neuman, D. R., Ash Disposal Pond Vegetation Die Off, unpubl. rep. prepared for the Montana Power Co., Montana State University, Bozeman, 1981.
79. Shacklette, H. T., Hamilton, J. C., Boerngen, J. G., and Bowes, J. M., Elemental Composition of Surficial Materials in the Conterminous United States, Prof. Paper 574-D, U.S. Geological Survey, Washington, D.C., 1971, D3.
80. Eaton, F. M., Deficiency, toxicity, and accumulation of boron in plants, *J. Agric. Res.*, 69, 237, 1944.
81. Gupta, U. C., Jame, Y. W., Campbell, C. A., Leyshon, A. J., and Nicholaichup, W., Boron toxicity and deficiency: a review, *Can. J. Soil Sci.*, 65, 381, 1985.
82. Severson, R. C. and Gough, L. P., Rehabilitation materials from surface coal mines in western U.S. I. Chemical characteristics of spoil and replaced cover soil, *Reclam. Reveg. Res.*, 2, 83, 1983.
83. Jones, H. E. and Scarseth, G. D., The calcium-boron balance in plants as related to boron needs, *Soil Sci.*, 56, 15, 1944.
84. Pakalne, D., Nollendorfa, A., and Upitis, V., Little investigated trace elements in *Chlorella* culture, *Biologija*, 11, 16, 1970, (in Russian).
85. Davies, B. E. and Roberts, L. J., Heavy metals in soils and radish in a mineralized limestone area of Wales, Great Britain, *Sci. Total Environ.*, 4, 249, 1975.
86. Wixson, B. G., The natural occurrence of cadmium in the environment in Proc. 1st Int. Cadium Conf., Metal Bull. Ltd., Surrey, U.K., 1977, 97.
87. Munshower, F. F., DePuit, E. J., and Neuman, D. R., *The Effects of Stack Emissions on the Range Resource in the Vicinity of Colstrip, Montana: Preoperational Description of the Range Ecosystem*, MT Agric. Exp. Sta. Res. Rep. 126, Montana State University, Bozeman, 1978, 15.
88. Munshower, F. F. and Neuman, D. R., *The Effects of Stack Emissions on the Range Resource in the Vicinity of Colstrip, Montana after Five Years of Operation of Power Plant Units III and IV*, MT Agric. Exp. Sta., Montana State University, Bozeman, 1987, 14.
89. Sherlock, J. C., Cadmium in foods and the diet, *Experientia*, 40, 152, 1984.
90. Shacklette, H. T., Cadmium in Plants, Bull. 1314-G, U.S. Geological Survey, Washington, D.C., 1972, 13.
91. Munshower, F. F., Cadmium accumulation in plants and animals of polluted and nonpolluted grasslands, *J. Environ. Qual.*, 6, 411, 1977.
92. National Environmental Research Center, Scientific and Technical Assessment Report on Cd, EPA-600/6–75–003, Nat. Environ. Res. Center, U.S. Environmental Protection Agency, Research Triangle Park, NC, 1975, 7–4.
93. Broyer, T. C., Carlton, A. B., Johnson, C. M., and Stout, P. R., Chlorine — a micronutrient element for higher plants, *Plant Physiol.*, 29, 526, 1954.

94. Gauch, H. C. and Wadleigh, C. H., Salt tolerance and chemical composition of Rhodes and Dallis grasses growing in sand cultures, *Bot. Gaz.*, 112, 259, 1951.
95. Dilley, D. R., Kenworthy, A. L., Benne, E. J., and Bass, S. T., Growth and nutrient absorption of apple, cherry, peach, and grape as influenced by various levels of chlorine and sulfate, *Proc. Am. Soc. Hortic. Sci.*, 72, 64, 1958.
96. Teater, R. W., Mederski, H. J., and Volk, G. W., Yield and mineral content of corn as affected by ammonium chloride fertilizer, *Agron. J.*, 52, 403, 1960.
97. Gough, L. P., Shacklette, H. T., and Case, A. A., Element Concentrations Toxic to Plants, Animals, and Man, Bull. 1466, U.S. Geological Survey, Washington, D.C., 1979, 17.
98. Turekian, K. K. and Wedepohl, K. H., Distribution of the elements in some major units of the earth's crust, *Geol. Soc. Am. Bull.*, 72, 175, 1961.
99. Adriano, D. C. and Doner, H. E., Bromine, chlorine, and fluorine, in *Methods of Soil Analysis, Part 2*, 2nd ed., Page, A. L., Ed., American Society of Agronomy and Soil Science Society of America, Madison, WI, 1982, 449.
100. Gupta, U. C., Copper in agricultural crops, in *Copper in the Environment, Part I*, Nriagu, J. O., Ed., John Wiley & Sons, New York, 1979, 255.
101. Woolhouse, H. W. and Walker, S., The physiological basis of copper toxicity and copper tolerance in higher plants, in *Copper in Soils and Plants*, Loneragan, J. F., Robson, A. D., and Graham, R. D., Eds., Academic Press, Sydney, 1981, 235.
102. Connor, J. J. and Shacklette, H. T., Background Geochemistry of Some Rocks, Soils, Plants, and Vegetables in the Conterminous United States, Prof. Paper 574-F, U.S. Geological Survey, Washington, D.C., 1975, 101.
103. Follett, R. H. and Lindsay, W. L., *Profile Distribution of Zinc, Iron, Manganese, and Copper in Colorado Soils*, Exp. Sta. Tech. Bull. 110, Colorado State University, Fort Collins, 1970, 32.
104. Munshower, F. F., DePuit, E. J., and Neuman, D. R., *The Effects of Stack Emissions on the Range Resource in the Vicinity of Colstrip, Montana: Preoperational Description of the Range Ecosystem*, MT Agric. Exp. Sta. Res. Rep. 126, Montana State University, Bozeman, 1978, 15.
105. Hamilton, J. W. and Gilbert, C. S., *Composition of Wyoming Range Plants and Soils*, Agric. Exp. Sta. Res. J. 55, University of Wyoming, Laramie, 1972, 7.
106. Munshsower, F. F., Neuman, D. R., and Rennick, R., *The Effects of Stack Emissions on the Range Resources in the Vicinity of Colstrip, Montana after Two Years of Operation of Colstrip Units 3 and 4*, Reclam. Res. Unit Publ. 8703, Montana State University, Bozeman, 1987, 14.
107. Agricultural Research Council, The Nutrient Requirements of Farm Livestock, No. 2, Her Majesty's Stationery Office, London, 1965, 101.
108. Subcommittee on Beef Cattle Nutrition and Committee on Animal Nutrition, *Nutrient Requirements of Beef Cattle*, 5th ed., National Academy of Sciences, Washington, D.C., 1976, 9.

109. Munshower, F. F. and Neuman, D. R., Elemental concentrations in native plant species growing on minesoils and native range, *Reclam. Rev.*, 3, 41, 1980.
110. Erdman, J. A., Ebens, R. J., and Case, A. A., Molybdenosis: a potential problem in ruminants grazing on coal mine spoils, *J. Range Manage.*, 31, 4, 1978.
111. Neuman, D. R. and Munshower, F. F., Copper and molybdenum uptake by forages grown on coal mine soils, *J. Range Manage.*, 37, 517, 1984.
112. Hamilton, J. W. and Gilbert, C. S., *Composition of Wyoming Range Plants and Soils*, Agric. Exp. Sta. Res. J. 55, University of Wyoming, Laramie, 1972, 7.
113. Munshower, F. F., DePuit, E. J., and Neuman, D. R., *The Effects of Stack Emissions on the Range Resource in the Vicinity of Colstrip, Montana: Pre-Operational Description of the Range Ecosystem*, MT Agric. Exp. Sta. Res. Rep. 126, Montana State University, Bozeman, 1978, 154.
114. Shacklette, H. T., Hamilton, J. C., Boerngen, J. G., and Bowes, J. M., Elemental Composition of Surficial Materials in the Conterminous United States, Prof. Paper 574-D, U.S. Geological Survey, Washington, D.C., 1971, D3.
115. Connor, J. J. and Shacklette, H. T., Background Geochemistry of Some Rocks, Soils, Plants, and Vegetables in the Conterminous United States, Prof. Paper 574-F, U.S. Geological Survey, Washington, D.C., 1975, 76.
116. Follett, R. H. and Lindsay, W. L., *Profile Distribution of Zinc, Iron, Manganese, and Copper in Colorado Soils*, Exp. Sta. Tech. Bull. 110, Colorado State University, Fort Collins, 1970, 30.
117. Munshower, F. F., DePuit, E. J., and Neuman, D. R., *The Effects of Stack Emissions on the Range Resource in the Vicinity of Colstrip, Montana: Pre-Operational Description of the Range Ecosystem*, MT Agric. Exp. Sta. Res. Rep. 126, Montana State University, Bozeman, 1978, 15.
118. Jenkins, D. W., Biological Monitoring of Toxic Trace Metals, Vol. 2. Part II, Off. Res. and Dev., EPA-600/3–80–091, U.S. Environmental Protection Agency, Las Vegas, NV, 1980, 725.
119. Munshsower, F. F., Neuman, D. R., and Rennick, R., *The Effects of Stack Emissions on the Range Resources in the Vicinity of Colstrip, Montana after Two Years of Operation of Colstrip Units 3 and 4*, Reclam. Res. Unit Publ. 8703, Montana State University, Bozeman, 1987, 14.
120. Maclean, A. J., Halstead, R. L., and Finn, B. J., Extractability of added lead in soils and its concentration in plants, *Can. J. Soil Sci.*, 49, 327, 1969,
121. Taylor, R. W. and Allinson, D. W., Influence of lead, cadmium, and nickel on the growth of *Medicago sativa*, *Plant Soil*, 60, 223, 1981.
122. Environmental Protection Agency, Assessment of the Toxicity of Arsenic, Cadmium, Lead, and Zinc in Soil, Plants, and Livestock in the Helena Valley of Montana, Contract No. 68–01–7251, Helena, MT, 1987, 117.
123. Davies, B. E., Beckett, P. H., and Wollan, E., Critical levels of twenty potentially toxic elements in young barley, *Plant Soil*, 49, 395, 1978.

124. Connor, J. J. and Shacklette, H. T., Background Geochemistry of Some Rocks, Soils, Plants, and Vegetables in the Conterminous United States, Prof. Paper 574-F, U.S. Geological Survey, Washington, D.C., 1975, 84.
125. Munshower, F. F., DePuit, E. J., and Neuman, D. R., *The Effects of Stack Emissions on the Range Resource in the Vicinity of Colstrip, Montana: Pre-Operational Description of the Range Ecosystem,* MT Agric. Exp. Sta. Res. Rep. 126, Montana State University, Bozeman, 1978, 15.
126. Clarkson, D. T. and Hanson, J. B., The mineral nutrition of higher plants, *Annu. Rev. Plant Physiol.*, 31, 239, 1980.
127. Hamilton, J. W. and Gilbert, C. S., *Composition of Wyoming Range Plants and Soils,* Agric. Exp. Sta. Res. J. 55, University of Wyoming, Laramie, 1972, 7.
128. Neuman, D. R. and Munshower, F. F., *Effects of Power Plant Emissions on the Range Resource in the Vicinity of Colstrip, Montana. Annual Report,* Reclam. Res. Unit, Montana State University, Bozeman, 1981, 33.
129. Gough, L. P., Shacklette, H. T., and Case, A. A., Element Concentrations Toxic to Plants, Animals, and Man, Bull. 1466, U.S. Geological Survey, Washington, D.C., 1979, 33.
130. Connor, J. J. and Shacklette, H. T., Background Geochemistry of Some Rocks, Soils, Plants, and Vegetables in the Conterminous United States, Prof. Paper 574-F, U.S. Geological Survey, Washington, D.C., 1975, 96.
131. Munshower, F. F., DePuit, E. J., and Neuman, D. R., *The Effects of Stack Emissions on the Range Resource in the Vicinity of Colstrip, Montana: Pre-operational Description of the Range Ecosystem,* MT Agric. Exp. Sta. Res. Rep. 126, Montana State University, Bozeman, 1978, 84.
132. Follett, R. H. and Lindsay, W. L., *Profile Distribution of Zinc, Iron, Manganese, and Copper in Colorado Soils,* Exp. Sta. Tech. Bull. 110, Colorado State University, Fort Collins, 1970, 31.
133. Munshower, F. F., DePuit, E. J., and Neuman, D. R., *The Effects of Stack Emissions on the Range Resource in the Vicinity of Colstrip, Montana: Pre-Operational Description of the Range Ecosystem,* MT Agric. Exp. Sta. Res. Rep. 126, Montana State University, Bozeman, 1978, 15.
134. Parker, M. B., Harris, H. B., Morris, H. D., and Perkins, G. F., Manganese toxicity of soybeans as related to soil and fertility treatments, *Agron. J.*, 61, 515, 1969.
135. Foy, D. D., Toxic Factors in Acid Soils of the Southwestern United States as Related to the Response of Alfalfa to Lime, Prod. Res. Rep. 80, U.S. Department of Agriculture, Washington, D.C., 1964, 14.
136. Follett, R. H. and Lindsay, W. L., *Profile Distribution of Zinc, Iron, Manganese, and Copper in Colorado Soils,* Exp. Sta. Tech. Bull. 110, Colorado State University, Fort Collins, 1970, 10.
137. Kabata-Pendias, A. and Pendias, H., *Trace Elements in Soils and Plants,* CRC Press, Boca Raton, FL, 1984, 229.
138. D'Itri, F. M., *The Environmental Mercury Problem,* CRC Press, Boca Raton, FL, 1972, 9.
139. Shacklette, H. T., Boerngen, J. G., and Turner, R. L., Mercury in the Environment — Surficial Materials of the Conterminous United States, Circ. 644, U.S. Geological Survey, Washington, D.C., 1971, 4.

140. Shacklette, H. T., Mercury content of plants, in Mercury in the Environment, Prof. Paper 713, U.S. Geological Survey, Washington, D.C., 1970, 35.
141. Shacklette, H. T., Hamilton, J. C., Boerngen, J. G., and Bowes, J. M., Elemental Composition of Surficial Materials in the Conterminous United States, Prof. Paper 574-D, U.S. Geological Survey, Washington, D.C., 1971, D3.
142. Grigg, J. L., Determination of available soil molybdenum, *N.Z. Soil News*, 3, 37, 1953.
143. Reisenauer, H. M., Molybdenum, in *Methods of Soil Analysis, Part 2*, Publ. No. 9, Black, C. A., Ed., American Society of Agronomy, Madison, WI, 1965, 1050.
144. Neuman, D. R., Shrack, J. L., and Gough, L. P., Copper and molybdenum, in *Reclaiming Mine Soils and Overburden in the Western United States*, Williams, D. R. and Schuman, G. E., Eds., Soil and Water Conservation Society, Ankeny, IA, 1987, 215.
145. Allaway, W. H., Prospectives on molybdenum in soils and plants, in *Molybdenum in the Environment. Vol. 2*, Chappell, W. R. and Peterson, K. K., Eds., Marcel Dekker, New York, 1977, 317.
146. Johnson, C. M., Pearson, G. A., and Stout, P. R., Molybdenun nutrition in crop plants. II. Plant and soil factors concerned with Mo deficiencies in crop plants, *Plant Soil*, 4, 178, 1952.
147. Munshower, F. F. and Neuman, D. R., Elemental concentrations in native plant species growing on minesoils and native range, *Reclam. Rev.*, 3, 41, 1980.
148. Neuman, D. R. and Munshower, F. F., Copper and molybdenum uptake by forages grown on coal mine soils, *J. Range Manage.*, 37, 517, 1984.
149. Gupta, U. C. and Lipsett, J., Molybdenum in soils, plants, an animals, *Adv. Agron.*, 34, 73, 1981.
150. Lesperance, A. L., Winning the battle against molybdenosis, *Nev. Ranch Home Rev.*, 3, 14, 1967.
151. VanBaalen, C. and O'Donnell, R., Isolation of a nickel- dependent blue-green algae, *J. Gen. Microbiol.*, 105, 259, 1978.
152. Nielsen, F. H. and Sauberlich, A. E., Evidence of a possible requirement for nickel by the chick, *Proc. Soc. Exp. Biol. Med.*, 134, 845, 1970.
153. Sundermann, R. W., Jr., Nomoto, S., Morang, R., Mechay, M. W., Burke, C. N., and Nielsen, S. W., Nickel deprivation in chicks, *J. Nutr.*, 102, 259, 1972.
154. Anke, M., Grun, M., Dittrich, G., Groppel, B., and Hennig, A., Low nickel rations for growth and reproduction in pigs, in *Trace Element Metabolism in Animals Vol. 2, Proc. 2nd Int. Symp. Trace Element Metabolism in Animals*, University Park Press, Baltimore, 1974, 715.
155. Connor, J. J. and Shacklette, H. T., Background Geochemistry of Some Rocks, Soils, Plants, and Vegetables in the Conterminous United States, Prof. Paper 574-F, U.S. Geological Survey, Washington, D.C., 1975, 109.

156. Shewry, P. R. and Peterson, S. J., Distribution of chromium and nickel in plants and soil from serpentine and other sites, *J. Appl. Ecol.*, 13, 195, 1976.
157. Slingsby, D. R. and Brown, D. H., Nickel in British serpentine soils, *J. Ecol.*, 65, 597, 1977.
158. Munshower, F. F., DePuit, E. J., and Neuman, D. R., *The Effects of Stack Emissions on the Range Resource in the Vicinity of Colstrip, Montana: Pre-Operational Description of the Range Ecosystem*, MT Agric. Exp. Sta. Res. Rep. 126, Montana State University, Bozeman, 1978, 15.
159. Mitchell, R. L., Cobalt and nickel in soils and plants, *Soil Sci.*, 60, 63, 1945.
160. Munshsower, F. F., Neuman, D. R., and Rennick, R. B., *The Effects of Stack Emissions on the Range Resources in the Vicinity of Colstrip, Montana after Two Years of Operation of Colstrip Units 3 and 4*, Reclam. Res. Unit Publ. 8703, Montana State University, Bozeman, 1987, 14.
161. Hunter, J. G. and Vergnano, O., Nickel toxicity in plants, *Ann. Appl. Biol.*, 39, 279, 1952.
162. Moore, T. R. and Zimmermann, R. C., Establishment of vegetation on serpentine asbestos mine wastes, southeastern Quebec, Canada, *J. Appl. Ecol.*, 14, 589, 1977.
163. Khalid, B. Y. and Tinsley, J., Some effects of nickel toxicity on rye grass, *Plant Soil*, 55, 139, 1980.
164. Adriano, D. C., *Trace Elements in the Terrestrial Environment*, Springer-Verlag, New York, 1986, 400.
165. Connor, J. J. and Shacklette, H. T., Background Geochemistry of Some Rocks, Soils, Plants, and Vegetables in the Conterminous United States, Prof. Paper 574-F, U.S. Geological Survey, Washington, D.C., 1975, 128.
166. Lakin, H. W., Selenium content of soils, in Selenium in Agriculture, Anderson, M. S., Ed., Agric. Res. Serv. Handbook 200, U.S. Department of Agriculture, Washington, D.C., 1961, 27.
167. Olson, O. E. and Moxon, A. L., The availability to crop plants of different forms of selenium in the soil, *Soil Sci.*, 47, 305, 1939.
168. Williams, C. and Thornton, I., The use of soil extractants to estimate plant-available molybdenum and selenium in potentially toxic soils, *Plant Soil*, 39, 149, 1973.
169. Trelease, S. F. and Beath, O. A., *Selenium*, Champlain Printers, Burlington, VT, 1949, 264.
170. Rosenfeld, I. and Beath, O. A., *Selenium. Geobotany, Biochemistry, Toxicity, and Nutrition*, Academic Press, New York, 1964, 321.
171. Olson, O. E. and Moxon, A. L., The availability, to crop plants, of different forms of selenium in the soil, *Soil Sci.*, 47, 305, 1939.
172. Carlstrom, M. G., Memorandum: Participating Laboratories, Third Round Robin Results, Montana Department of State Lands, Helena, 1985, 1.
173. Severson, R. C. and Fisher, S. E., Jr., Results of the First Western Task Force Round Robin Soil and Overburden Analysis Program, Open-file Rep. 85–220, U.S. Geological Survey, Denver, CO, 1985, 5.

174. Soltanpour, P. N. and Workman, S. M., Use of ammonium bicarbonate-DTPA soil test to assess availability and toxicity of selenium to alfalfa plants, *Commun. Soil Sci. Plant Anal.*, 11, 1147, 1980.
175. Olson, O. E. and Moxon, A. L., The availability to crop plants of different forms of selenium in the soil, *Soil Sci.*, 47, 305, 1939.
176. Olson, O. E., Whitehead, E. I., and Moxon, A. L., Occurrence of soluble selenium in soils and its availability to plants, *Soil Sci.*, 54, 47, 1942.
177. Lakin, H. W., Selenium accumulation in soils and its absorption by plants and animals, *Geol. Soc. Am. Bull.*, 83, 181, 1972.
178. Prodgers, R. A. and Munshower, F. F., AB-DTPA extractable soil selenium as a predictor of seleniferous vegetation, in Proc. 1990 Billings Land Reclam. Symp. Selenium in Arid and Semiarid Environments, Western U.S., U.S. G.S. Circ. 1064, U.S. Geological Survey, Denver, 1991, 65.
179. Jenkins, D. W., Biological Monitoring of Toxic Trace Metals. Vol. 2. Part III, EPA-600/3–80–092, Off. of Res. and Dev., U.S. Environmental Protection Agency, Las Vegas, NV, 1980, 1116.
180. Carter, D. L., *Plan Ahead to Prevent Livestock Losses From White Muscle Disease*, Curr. Inf. Ser. No. 84., Agric. Exp. Sta., University of Idaho, Moscow, 1968, 1.
181. Graham, D. R., *Selenium and Livestock Nutrition in Montana*, Agric. Exp. Sta. Res. Rep. 31, Montana State University, Bozeman, 1973, 2.
182. Schwarz, K. and Foltz, C. M., Selenium as an integral part of factor 3 against dietary necrotic liver degeneration, *J. Am. Chem. Soc.*, 79, 3292, 1957.
183. Committee on Medical and Biologic Effects of Environmental Pollutants, *Selenium*, National Academy of Sciences, Washington, D.C., 1976, 111.
184. Trelease, S. F. and Beath, O. A., *Selenium*, Champlain Printers, Burlington, VT, 1949, 123.
185. Beath, O. A., Gilbert, C. S., and Eppson, H. F., The use of indicator plants in locating seleniferous areas in western United States, *Am. J. Bot.*, 26, 257, 1939.
186. Whanger, P. D., Bioenvironmental Impact of Selenium, Natl. Ecol. Res. Lab., U.S. Environmental Protection Agency, Corvallis, OR, 1974, 61.
187. Brownell, P. F. and Wood, J. G., Sodium as an essential micronutrient element for *Atriplex visicaria* Heward, *Nature*, 179, 635, 1957.
188. Brownell, P. F. and Jackson, M. E., Changes during recovery from sodium deficiency in *Atriplex*, *Plant Physiol.*, 41, 617, 1966.
189. Williams, M. C., Effect of sodium and potassium salts on growth and oxalate content of halogeton, *Plant Physiol.*, 35, 500, 1960.
190. Connor, J. J. and Shacklette, H. T., Background Geochemistry of Some Rocks, Soils, Plants, and Vegetables in the Conterminous United States, Prof. Paper 574-F, U.S. Geological Survey, Washington, D.C., 1975, 135.
191. Clarkson, D. T. and Hanson, J. B., The mineral nutrition of higher plants, *Annu. Rev. Plant Physiol.*, 31, 239, 1980.

192. Connor, J. J. and Shacklette, H. T., Background Geochemistry of Some Rocks, Soils, Plants, and Vegetables in the Conterminous United States, Prof. Paper 574-F, U.S. Geological Survey, Washington, D.C., 1975, 161.
193. Follett, R. H. and Lindsay, W. L., *Profile Distribution of Zinc, Iron, Manganese, and Copper in Colorado Soils*, Exp. Sta. Tech. Bull. 110, Colorado State University, Fort Collins, 1970, 29.
194. Munshower, F. F., DePuit, E. J., and Neuman, D. R., *The Effects of Stack Emissions on the Range Resource in the Vicinity of Colstrip, Montana: Pre-Operational Description of the Range Ecosystem*, Agric. Exp. Sta. Res. Rep. 126, Montana State University, Bozeman, 1978, 15.
195. Follett, R. H. and Lindsay, W. L., *Profile Distribution of Zinc, Iron, Manganese, and Copper in Colorado Soils*, Exp. Sta. Tech. Bull. 110, Colorado State University, Fort Collins, 1970, 10.
196. Brown, A. L., Quick, J., and Eddings, J. L., A comparison of analytical methods for soil zinc, *Soil Sci. Soc. Am. Proc.*, 35, 105, 1971.
197. Yadav, O. P. and Shukla, U. C., Effect of zinc on nodulation and nitrogen fixation in chickpea (*Cicer arietinum* L.), *J. Agric. Sci. Camb.*, 101, 559, 1983.
198. Environmental Protection Agency, Assessment of the Toxicity of Arsenic, Cadmium, Lead, and Zinc in Soil, Plants, and Livestock in the Helena Valley of Montana, Contract No. 68–01–7251, Helena, MT, 1987, 118.
199. Yadav, O. P. and Shukla, U. C., Effect of zinc on nodulation and nitrogen fixation in chickpea (*Cicer arietinum* L.), *J. Agric. Sci. Camb.*, 101, 559, 1983.
200. Hamilton, J. W. and Gilbert, C. S., *Composition of Wyoming Range Plants and Soils*, Agric. Exp. Sta. Res. J. 55, University of Wyoming, Laramie, 1972, 9.
201. Munshsower, F. F., Neuman, D. R., and Rennick, R. B., *The Effects of Stack Emissions on the Range Resources in the Vicinity of Colstrip, Montana after Two Years of Operation of Colstrip Units 3 and 4*, Reclam. Res. Unit Publ. 8703, Montana State University, Bozeman, 1987, 14.
202. Gough, L. P., Shacklette, H. T., and Case, A. A., Element Concentrations Toxic to Plants, Animals, and Man, U.S.G.S. Bull. 1466, U.S. Geological Survey, Washington, D.C., 1979, 56.
203. Kabata-Pendias, A. and Pendias, H., *Trace Elements in Soils and Plants*, CRC Press, Boca Raton, FL, 1984, 107.

3

Site Preparation

Although many people assume that revegetation is a farming activity, such is not the case. While modern methods of disturbed land reclamation evolved from the agronomic practices of early restoration efforts and is still influenced by them, revegetation has progressed far beyond farming since early rehabilitation efforts. In this chapter we will discuss land reshaping activities as well as those standard farming practices that are an integral part of today's revegetation programs. We will also examine several components of these programs that are not considered agronomic practices. Subsequent chapters evaluate the contributions made by numerous revegetation specialists to adapt these practices to land rehabilitation. These individuals include mine company reclamation scientists, consultants, researchers, seed suppliers, range and forestry experts, botanists, and others who have developed new techniques to enhance the germination, establishment, and growth of vegetation on disturbed lands and to restore the appearance of those lands after revegetation.

Revegetation technology today is strongly influenced by processes such as recontouring, topsoil application, surface manipulation, and seeding or planting of selected plant populations. The importance of each of these activities cannot be over emphasized. However, on some disturbances other practices are incorporated into the revegetation phase to ensure both short- and long-term plant community development. These additional practices may range from simple topsoil mulching, through subsoiling over sodic materials, to lime amendment of pyritic overburden before topsoil application.

Revegetation experts are tempted to incorporate into reclamation programs every biologically sound improvement technique known to rehabilitation science. Individuals must realize, however, that each activity incorporated into the program requires capital investment and no restoration program has an unlimited budget. Each reclamation activity must meet some criteria of usefulness to the rehabilitation effort, but clarification of these criteria is difficult. In a market dominated by profit margins, useful criteria would be identified by this factor alone, but our society is only partially governed by the profit factor. Other expressions of human interest enter into the reclamation equation. These factors include laws and regulations, aesthetics, and public opinion. They blunt the cold edge of the profit margin.

Restoration efforts on large-scale disturbances are usually made in response to legal requirements. Legislated reclamation does not need to meet the test of economic feasibility (will reclamation pay for itself?), or economic efficiency (will it yield a higher return than other potential uses of the capital?). However, the project should be cost effective. That is, the revegetation program should produce a specified result at a minimum expenditure of capital.[1] The product of the revegetation specialist is plant growth on the degraded landscape. In one example, this may equate to *just turn it green;* in another example, restoration may demand a plant community equal to the potential of the site prior to disturbance. In all instances, however, the final vegetation cover should be developed at minimum cost.

Soil and overburden movement are the most expensive parts of disturbed land rehabilitation. Thorough planning and coordinated soil/overburden handling during disturbance can reduce the cost of this very expensive stage of degraded land reclamation. The costs of seedbed preparation, seed, soil amendments, and plants are small compared to recontouring, subsoiling, and topsoiling. Revegetation failure caused by improper seedbed preparation or poor quality seed introduces major costs if scrappers and dozers must be mobilized on the disturbance a second time because of erosion encouraged by poor plant performance.

RECONTOURING

One of the primary consideration in land reclamation is erosion control. Landforms remaining after area mining, highway construction, and many other types of disturbance are often steep and have shapes that do not conform to predisturbance drainage patterns. Such degraded lands support little vegetation, are unstable, and contribute sediment to local streams or rivers. The objectives of land recontouring, therefore, are slope stability, reconstruction of the original drainage basin topography, and satisfactory surfaces for plant establishment. All of these processes help reduce soil erosion.

In semiarid and arid environments, water is the major factor limiting plant growth. Restoring plant production in such areas requires the construction of landforms and soils that conserve the water resource. Increasing the length and steepness of a slope increases runoff and decreases infiltration. The resultant loss of soil moisture decreases the probability of vegetation establishment and growth. Plant performance is rarely acceptable on disturbed slopes greater than 2:1.[2]

The lack of a general guide for recontoured surfaces plagued early rehabilitation efforts and is still a problem with hardrock mines. The modification of landforms to approximate original contour (AOC) became a serious component of rehabilitation programs in the coal fields of the United States only after passage of mine reclamation laws in the 1970s. After the phrase AOC was incorporated into the Surface Mining Control and Reclamation Act (SMCRA) in 1977, it slowly became accepted by all responsible coal mining companies. It is not practiced in metal mining for a number of reasons. These mines are usually underground, in mountainous terrain with very steep slopes, or are large open pits which do not lend themselves to backfilling and grading during or after mining. However, even in these types of disturbances the return to AOC or slopes less steep than the angle of repose reduces environmental impact.

During the planning phase for land disturbances, postdisturbance site topography can be mapped. Computer software can draft landscapes to conform to predefined slope and exposure parameters and duplicate predisturbance landforms. If excess spoil exists, as it does in many eastern coal mines, it can be placed in properly constructed head of hollow or valley fills. Because of the thickness of western coal seams and their proximity to the soil surface, excess spoil is rare in this region. Stripmines usually can be returned to AOC in the West without resorting to excess spoil disposal techniques. Indeed, a lack of spoil is more likely to be found in western North American coal mines than excess spoil.

Spoils recontouring must be carried out with the realization that the upper layer of these materials constitutes the rootzone of the future plant community growing in the coversoil. If any spoil material reveals potential or existing phytotoxic properties it must be buried under a nontoxic subsoil layer. Indirect evidence from studies of rooting zones in the Great Plains has shown that over 120 cm (4 ft) and possibly as much as 245 cm (8 ft) of nontoxic material are required to support a diverse, native grassland community.[3–5] Numerous examples from mesic to semiarid zones have demonstrated that lesser amounts of soil (e.g., 45 cm [18 in.]) are sufficient to support a simple stand of shallow-rooted, drought-tolerant grasses, but there are no examples of the restoration of complex plant communities on such thin soils. Toxic spoil should, therefore, be placed at depths greater than 120 cm and probably as deep as 200 to 250 cm (6.5 to 8.2 ft) to ensure that an adequate depth of soil is available for the development of healthy root systems of diverse plant communities.

Recontouring should tie the degraded site into adjoining undisturbed terrain (Figure 3). Vegetation germination and establishment

Figure 3 Recontoured minesoil and adjacent native rangeland.

should further obscure the boundary between disturbed and native plant communities. Recontoured slopes should flow smoothly into the undisturbed landscape and the reconstructed drainage pattern should permit the uninhibited flow of water from rehabilitated to untouched stream channels. The reconstructed landscape and the surrounding ecosystem should be a smooth, continuous sequence of hill slopes and valley floors.

Slope Reconstruction

The stability of any slope, including undisturbed hillslopes, is relative. Slopes that are not eroding at an accelerated rate are in a state of temporary equilibrium. The equilibrium results from the interaction of the climate of the area with several factors:

1. Slope geomorphology — including shape, length, and gradient of the landform
2. Properties of the materials that make up the slope — including infiltration rate, porosity, bulk density, etc.
3. Vegetation growing on the slope

When a hillside is disturbed, the level of degradation varies with the intensity of the destruction. The surface soil layers may be impacted but intact, as after a fire, or they may have been picked up, mixed to some degree, and placed in storage for months to years. The underlying materials may be intact as after a fire, or they may have been violently disturbed as by mining. Mining is one of the types of disturbances that completely destroys the geologic structure of a slope, many of the properties of the materials composing the slope, and the vegetation covering on the slope.

Any lithologic control of erosion that existed on premined slopes is destroyed by mining and cannot be recovered by reclamation processes. The geomorphology of hillslopes is also destroyed by mining, but postdisturbance geomorphology is an artifact of spoil handling.[6] Slope shape, length, and gradient are subject to modification during the reclamation process.

The loss of slope structure, physical properties, and vegetation covering necessitates that hillslopes constructed after mining evolve through plant growth and soil erosion toward a new equilibrium with the climate of the area. Unless other alternatives are implemented, erosion will occur on the replacement hillslope even after proper recontouring.

Major differences have been reported between the amount and types of erosion on undisturbed and mined land hillslopes. In at least one study in Colorado, rilling and gully formation were common on mined lands while there was only minor sheet erosion on unmined control slopes.[7] Increased erosion and the different types of erosion seen on degraded slopes were assumed to be results of the destruction of the equilibrium that existed on the slope prior to disturbance. In these same studies, infiltration rates on undisturbed soils were an order of magnitude greater than infiltration rates on topsoiled minesoils.[8] However, a similar investigation on relatively flat minesoils in Montana did not find any differences between infiltration rates on native prairie and topsoiled minesoils.[9] There is no agreement as to the impact of mining and rehabilitation on infiltration other than the consensus that compaction occurs in the reconstruction of minesoil profiles. In the Colorado investigations, porosity was reported to be lower and bulk density higher on the mined land slopes. These two factors probably contributed to the differences between infiltration rates reported on mined and undisturbed hillslopes.

Many undisturbed slopes approach the angle of repose for waste rock dumps (<2.0:1). These hillsides often appear to be free of erosion but this stability is a result of slope lithology, shape, length, gradient, and surface vegetation. Underlying geologic materials are composed of consolidated rock or lightly weathered materials, and soil is held tightly to the near-surface parent material by a maze of plant roots intertwined with the uneven surface and broken segments of the parent rock. The mature canopy of grasses, forbs, shrubs, and trees sends its roots deep into the hillside. These plant roots tenaciously grip the soil and rocks buried in the hillslope and near surface soil. Stems and leaves break the force of raindrop impact on the soil surface. Vegetation and the coarse texture of the rootzone materials ensure rapid and deep infiltration of precipitation. These properties are strong slope-stabilizing influences.

After blasting and movement, both the consolidated subsurface and plant roots are lacking in the waste material. Erosion is exaggerated by this loss. The ecological processes that led to the formation of a stable plant community on the slope prior to disturbance must be repeated before vegetation can exert the same erosion-controlling force on the slope after degradation that it exerted before disturbance. To achieve the level of soil stability found on hillslopes covered with mature plant communities may take two or three growing seasons after a fire, unless the fire was unusually severe. To attain the same level of stability after mining may take decades to centuries, varying with the geographic location of the disturbance and climatic factors.

The crux of the problem of violently disturbed slope stabilization is to stimulate plant growth and encourage development of plant roots before unacceptable erosion or mass slumping occurs.

It has been pointed out that in semiarid and arid areas available soil moisture is the main limiting factor to vegetation growth. Survival and growth of plants on any hillslope is reduced compared to that measured on more level soils because water loss is exacerbated by the gradient of the slope. Soil infiltration rates, water-holding capacity of soil and subsoil, and runoff contribute to the water budget on a slope. Plant germination and establishment on steep slopes are directly related to this budget. The rate of development of plants on a slope are subjected to the normal aridity of the area plus a moisture deficit aggravated by accelerated runoff.

During reconstruction, slope shape can be manipulated to reduce water loss and soil erosion. Constructed slopes may be concave, convex, or straight, and may be simple or grouped into complex surfaces. Concave slopes are more erosion resistant and stable than straight or convex slopes and complex slopes are more stable than a concave, convex, or straight slope. Within the limits of the size of the disturbance, slope length may also be adjusted during the planning and recontouring phases of reclamation. A group of researchers[10] reported that soil loss increased as a power function of slope length, in which the value of the exponent was determined by soil erosion factors, but was usually less than one. They further reported that erosion increased rapidly as slope gradient increased.

Earth moving and mining rely heavily on the use of large equipment — draglines, scrappers, and dozers. The use of this machinery is increasingly limited by slopes, especially as the slope approaches or exceeds 3:1. Wheeled vehicles are rarely used on slopes steeper than this, and operation of large equipment on the contour becomes increasingly dangerous as the slope steepens. Because most final soil rehabilitation practices including seedbed preparation, should be carried out on the contour, this slope (3:1) becomes a practical as well as legal reclamation limit. The Forest Service recommends a slope limit of 2.5:1 for waste dumps because of safety hazards posed by equipment operation, the movement of seed and fertilizer downslope by erosional forces, and the poor operation of seed drills or cultipackers on such slopes.[11]

Coal stripmine regulations require that all highwalls be reduced to slopes of 3:1 or less to retard erosion and permit the use of heavy equipment on the contour. The maximum slope permitted by coal mine regulations on the interior of surface coal mines is 5:1 (Figure 4).[12] This gradient (5:1) contributes to stability and reduced erosion but presents some shortcomings. It prevents the creation of diverse habitats needed to ensure the survival of different plants and animals on the site after reclamation.

Regulations governing hardrock mines and other disturbances place few limitations on slope steepness or length. Some restrictions on the slope of waste rock dumps may be incorporated into operating permits issued by state regulatory agencies, but the angle of repose often becomes the slope on which revegetation must be attempted. Excessive cost for the reduction of long steep slopes and operator safety become limiting factors, and reclamation suffers. Because topsoil is poorly distributed (if even utilized) and seed and fertilizer distribution is poor, weedy species or barren mineral soils dominate the landscape. If slopes are revegetated, the threats of erosion or mass slumping persist for many years. Successful, long-term revegetation and stabilization of slopes at or near the angle of repose has not been demonstrated.

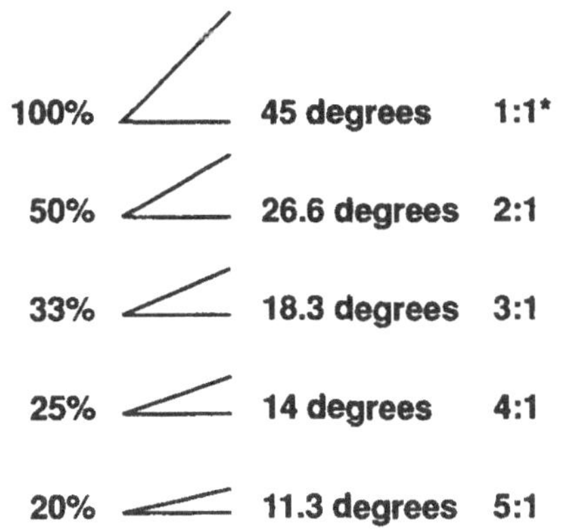

Figure 4 Slopes, angles, and grades. Asterisk indicates horizontal: vertical distance.

Steep slope revegetation is the emphasis of at least one demonstration currently underway. At the Golden Sunlight Mine in Montana, slopes of 2:1, 2.5:1, and 3:1 (50, 40, and 33%) were subjected to a number of treatments. The slopes were capped with oxidized rock, topsoiled, pitted with large depressions, broadcast seeded, and hydromulched with a tackifier. In the second growing season the slopes had remained stable and erosion was controlled.

Slope length and gradients remain the major handicap to successful revegetation of non-coal mines in almost any terrain. The preferred rehabilitated hillslope is concave or complex in profile, has a low gradient, and a short length. It maintains the high infiltration rate of the undisturbed slope and supports a vigorous stand of rapidly growing vegetation. When space restrictions, terrain, or other factors prevent the construction of such slopes, the incorporation of lifts or terraces with short steep slopes permits the establishment of selected vegetation on the flat terraces.[13] Short, steep slopes are less likely to exhibit severe erosion than long but somewhat less-steep slopes. The stability of the site, and accessibility afforded livestock and wildlife by terraces, increases the value of such reclamation practices.

Integrity of Drainage Basins

A drainage basin may range from a few small hillslopes and a short stretch of a first-order stream to an entire watershed of hills and valleys containing second- or third-order channels. In the undisturbed environment these landforms are relatively stable as a result of thousands of years of natural weathering, erosion, deposition, and plant growth. Like slopes, entire watersheds are in approximate equilibrium with the geologic, climatic, vegetative, and hydrologic forces operating on them. When the drainage basin or a segment of the basin is disrupted by any type of disturbance, this equilibrium is destroyed. During the reclamation phase, earth moving and recontouring are designed to create a new drainage basin that is in equilibrium with the surrounding ecosystem and climate of the region. Reclamation attempts to duplicate many years of interaction among these forces (geologic, climatic, vegetative, and hydrologic) but it cannot duplicate the geologic structure of the undisturbed site. The geologic structure and the vegetation of the mined area have been permanently altered. Because of these alterations, the geomorphic conditions of the reclaimed site are not in equilibrium with the climatic, vegetative, and hydrologic forces of the new drainage.[14] There must be some compensation for the loss of lithologic or geologic control of site stability. This compensation may take the form of reduced slope gradient, slope length, or rapid vegetation growth on the disturbance.

If reclamation specialists are aware of conditions that decrease drainage basin stability, the postmine landscape can be constructed to minimize those conditions. Slope, valley floor, and stream channel characteristics need to be incorporated into disturbed land reclamation plans.

The characteristics that contribute to erosion resistance on slopes have been discussed. Additional parameters that contribute to basin-wide stability include drainage channel density and pattern as well as stream channel characteristics. It has been suggested that first-order streams (the smallest channels in a watershed) not be constructed during reclamation of the drainage basin. Instead, they should be allowed to regenerate under natural processes because sediment yields are less from naturally eroded channels than from man-made channels.[15]

Stream channel stability is an indication of the adequacy of drainage basin reclamation.[16] The stream sits in the valley floor and the geomorphology of the flood plain and the slopes of the drainage directly influence the stability of the channel. The variables contributing to the morphology of the stream channel include gradient, width, bed material, and runoff, among others. While these are violently disturbed by mining they are all subject to manipulation to some degree during the reclamation process. Unfortunately, reclamation specialists tend to construct narrow, steep-sided drainage basins rather than broad valley bottoms with short, gentle, upland slopes.

Cross-sectional area, rate of fall, length, and vegetation control the amount of water that the drainage channel can hold and the rate of movement of that water through the channel. Major precipitation or snow melt events and geomorphic stability determine the sinuosity of the channel within the flood plain. A longer channel (greater sinuosity) reduces the slope of the drainage channel, increases the amount of water the channel can safely transport, and reduces the velocity of water flowing through the channel.[17] Grasses, shrubs, and trees growing in the drainage channel filter sediment, slow overland flow, increase sediment deposition, and reduce the erosive power of water. The ability of the entire watershed to absorb the erosive force of major runoff events without large-scale changes in the ecosystem is reinforced by yearly cycles of plant growth and changes in basin structure caused by this vegetation.

Reconstructed stream channels do not have the protective covering of perennial grasses, forbs, shrubs, or trees found in undisturbed riparian habitats. Because vegetation cover is lacking, the new drainage system is unstable and sensitive to large-scale perturbations from only moderate precipitation or runoff. Therefore, it is imperative that, at a minimum, reconstructed drainages on large-scale land disturbances such as area mines duplicate the original watershed pattern. Hillslopes, stream channel fall, and sinuosity must all mirror the predisturbance character of the ecosystem. Vegetation must also be seeded or planted as soon as possible to begin the successional sequence that contributes a greater degree of stability to the disturbed ecosystem.

Despite recontouring that duplicates predisturbance landforms, the loss of geologic and vegetational components of the drainage basin reduce the stability of the watershed, or the ability of the system to resist erosion. It may be prudent in semiarid and arid areas to over-design drainage basins to increase their stability while vegetation is developing. That is, design the hillslopes, valley floor, and stream channel to achieve a stability greater than that which existed prior to disturbance. As was pointed out in the previous section, infiltration rates on reclaimed slopes are reduced compared to predisturbance infiltration rates. Runoff, therefore, is augmented by the amount of water that the reclaimed soil cannot absorb. This increases the water load on the channel and hence the probability of bank erosion and increased sediment leaving the disturbed area. Vegetation on slopes and valley floor does not possess the maturity of vegetation found in the undisturbed environment and can not impart to the drainage basin the stability known in the premine ecosystem. If slope lengths and angles are reduced and channel gradient decreased, the newly constructed watershed may be stable enough without mature stands of vegetation to resist excessive erosion.

TOPSOILING

As used in SMCRA and most rehabilitation plans, topsoil is a general term. It applies either (1) to the soil surface, usually to the plow depth or the minimum depth that a scrapper operator can successfully lift from the soil, or (2) to the depth of material that expresses good plant growth-supporting characteristics. This may not be the surface, or A horizon as defined by the soil scientist. At many disturbances in arid regions there is no A horizon, or if there is an A horizon it may only extend for a few centimeters beneath the soil surface. What is called topsoil at these sites may be an A horizon, a combined A and B, only a B horizon or, in some cases, combined B and C horizons. At some disturbances a paucity of good surface soil necessitates the salvage of any material revealing characteristics that are conducive to plant growth. These materials may include small areas of deep alluvial deposits along drainage channels or deeper strata in the overburden. The term coversoil is more appropriate for this material than topsoil.

In any case, coversoil or topsoil has several properties that make it vastly superior to spoil as a plant growth medium. Surface soils or overburden materials salvaged for use as coversoil usually reveal:

1. Better physical properties, including soil structure, cation exchange capacity, and infiltration rate

2. More favorable chemical properties, including pH, and mineral concentrations

Surface soils alone reveal:

1. More organic matter
2. A greater number of microorganisms, including bacteria, fungi (including mycorrhizae), algae, and protozoans
3. More numerous insects, other arthropods, and worms
4. Living and intact plant propagules, including seeds, plant fragments, and whole plants
5. Larger and biologically active plant nutrient pools

Research has shown that even a few centimeters of surficial soil improves infiltration.[18] Seed germination and seedling establishment are also better in topsoiled materials. Studies in North Dakota have revealed that topsoiled sites support greater yields than nontopsoiled areas.[19]

The stripping, storage, and/or respreading of surface soils on disturbed surfaces are widely practiced by the surface mining industry. They did not become widely practiced, however, until passage of mine reclamation laws in the 1970s. This practice is expensive and is still resisted by part of the mining industry, state highway departments, and contractors responsible for revegetation of rights-of-way or other disturbances not closely regulated by state or federal agencies. The underground metal mining industry, for example, has resisted efforts to require topsoiling waste rock dumps. Given the type of terrain in which most hardrock mining takes place this is to be expected. Soils are very shallow over bedrock at the vast majority of hardrock mines and the industry does not consider topsoil stripping a worthy expenditure of money at these sites. Where a topsoil resource exists, however, the stripping and storage until after mining or the direct application of the coversoil to a disturbed surface facilitates the establishment and growth of vegetation.

Federal coal mine regulations define topsoil or coversoil. Typical limits to parameters that may reduce the usefulness of this material are shown in Table 13. This table represents only one of numerous regulatory attempts to define what constitutes an adequate surficial soil resource. Most regulations are better written and more realistic today than in 1977. When topsoil characteristics exceed any of the limits for the defined physical or chemical properties it may not be used on coal minesoils. Substitute materials must be located and used as the surficial soil layer. These substitute materials may be selected from deep in the overburden. Regardless of the plant growth-enhancing characteristics of overburden layers used as coversoils, the

TABLE 13 Suspect Levels of Various Constituents of Soils and Overburden[20]

Parameter	Suspect Level
Physical	
pH	8.8–9.0 or above
Conductivity	4 to 6 mmhos/cm or more
SAR	12 or higher
Texture	40% clay, or loamy sand and sand
Chemical (μg/g)	
Al	Not noted
B	8
Cd	0.1–1.0
Cu	40
Pb	pH <6; 10–15 pH >6; 15–20
Mn	60
Se	2.0
Zn	40

lack of organic matter and nutrients necessitates the addition of some form of organic amendment and fertilizer when these materials are placed on the soil surface. Organic matter and nutrients stimulate the establishment of normal biological cycles in these materials.

While not as valuable as directly hauled surface soil, stored topsoil is definitely better than substitute materials or no surface soil.[21] Stored topsoil tends to show reduced biological activity, in part due to a loss of mycorrhizae, bacteria, and invertebrates.[22,23] Stored surface soils also reveal a loss of organic matter,[24,25] adverse modification of structural qualities that contribute to better seedling establishment,[26] a reduced number of plant propagules,[27,28] and fragmented nutrient cycles.[29] Therefore, organic amendment and fertilization of topsoil that has been in storage for several years is necessary to ensure a rapid buildup of microbial populations and initiation of nutrient cycling.

Topsoil that has not been placed in storage but has been stripped and reapplied to a subsoil or spoil in one operation (direct-haul topsoil) retains intact nutrient cycles that are so important to the biological functions of the minesoil. Also present in freshly stripped topsoil are mycorrhizae in species and in numbers comparable to premine populations. The value of complete nutrient cycles and mycorrhizae to grassland and forest rehabilitation and succession are becoming clearer each year. In Yellowstone National Park, research has shown that topsoil stripped and replaced within 90 days retains viable populations of mycorrhizae, but topsoil stored over winter loses most of its mycorrhizal propagules.[30]

The depth of topsoil to be applied to overburden has been investigated by a number of researchers.[31,32] All agree that the coversoil depth necessary to achieve predisturbance plant production and cover varies with the properties of the underlying materials. Barth and Martin's study[32] is typical of most of this research. They reported that generic spoil — defined as nonalkaline nonsaline loams with SARs below 7 — showed an increase in plant production up to a topsoil depth of approximately 50 cm (20 in.). Sodic spoil, or material with SARs of 26 or higher, required a minimum of 70 cm (28 in.) of nonsodic topsoil to reach maximum cool season grass production and prevent sodium movement into surface horizons. Topsoiled acid spoils — defined as spoils exhibiting pH values between 3.6 and 4.3 — revealed increased grass production up to the maximum topsoil depth used in these studies (152 cm [60 in.]).

The movement of sodium into nonsodic topsoil has been discussed in several studies.[33-35] Research has shown that if the surface covering is thick enough, and of appropriate texture and clay mineralogy, sodium movement into the surface horizon approaches an equilibrium and sodication of the surface may be prevented.[36] Salts apparently migrate toward the soil surface during dry periods and are leached down by water infiltrating the profile during more mesic periods. Hopefully, sufficient nonsodic coversoil will be applied so that a dynamic equilibrium can be maintained between the salts and the climatic regime of the site and a nonsodic surface soil can be maintained.

Root development in spoils is usually limited, but in older nontopsoiled minesoils, roots have been reported to over 150 cm (60 in.).[37] While most grassland roots are concentrated in the surficial 50 cm (20 in.) of soil, roots penetrating much deeper may be necessary for plant survival during prolonged drought; 50 cm of good quality surface material and subsoil are a minimum to ensure an adequate soil depth for short-term root development and survival of rangeland vegetation. Research on root development of grassland plant communities[38-40] indicates that development of diverse plant communities may require deeper layers of materials conducive to root growth.

SURFACE MANIPULATION

To guarantee maximum germination and plant establishment, the surface of disturbed lands must be prepared to receive seed. It is rare that a site with or without topsoil is ready for direct drilling or broadcasting of seed without some sort of preparatory treatment. Most types of surface manipulation are standard agricultural techniques such as plowing, chiselling, disking, or harrowing. Ripping and gouging, including the digging of dozer basins, may also be part of the surface manipulative program, but they are not normally considered standard agricultural techniques.

All of these activities require large vehicles to pull equipment over or through the disturbed soil. To reduce the probability of erosion they should be practiced on the contour. As was pointed out earlier (see section on Slope Reconstruction) the implementation of these rehabilitation techniques with heavy equipment is increasingly limited by steep slopes.

Standard Farming Techniques

Plowing is primarily used to break up compacted surface soils, mix amendments with the surficial soil layer (Figure 5), and kill unwanted vegetation. The standard moldboard plow is rarely used on large-scale disturbances today because degraded sites are usually seeded during the first season conducive to seed germination after topsoil application. Such soils will not have had time to compact or grow a weed crop prior to seeding. Chiselling is often substituted for plowing (Figure 6). A chisel can break up a soil crust, incorporate fertilizer, and kill shallow-rooted weeds that may have become established. Disking may also be substituted for plowing (Figure 7). Both disking

Figure 5 Plowing a minesoil with an enlarged plow.

Figure 6 A chisel on a minesoil.

and chiselling are usually followed by harrowing (Figure 8) to break up small clods and prepare a final seedbed for drill seeding. If broadcast seeding is to be employed, either chiselling or disking may be the final seedbed preparatory activity, since a rougher seedbed is more desirable for broadcast seeding than for drill seeding.

Figure 7 Agricultural disk at a coal mine.

Figure 8 Chisel and harrow in final pass over a minesoil prior to drill seeding.

Final tillage should prepare a seedbed that meets several plant growth requirements. The bed must be loose enough for good water infiltration yet firm enough to support mechanical seeders and provide good soil-seed contact. Roots must contact the soil within a few millimeters after initiating growth to ensure an adequate water supply for the emerging seedling. The seed root must be enclosed within a moist atmosphere to prevent desiccation of the root hairs. At the same time the seedbed must be either smooth enough for a drill seeder to place the seeds at a uniform depth in the soil, or rough enough to provide the multiple niches necessary for good seed distribution by broadcast seeders. The chisel or disk followed by a spring-toothed harrow and some type of cultipacker usually prepares an adequate bed for drill seeding. A chisel or disk operated to disturb only 5 to 10 cm (2 to 4 in.) provides a satisfactory seedbed for most broadcast seeding.

Nontypical Seedbed Preparation

Of the nonconventional activities utilized in the preparation of seed beds on disturbed soils, ripping is probably the most often encountered technique. It is the practice of pulling a steel tooth or shank through the soil, subsoil, or overburden. Ripper teeth (Figure 9) are commonly over 45 cm (18 in.) long. The distance between ripper teeth is generally equal to the length of the teeth.[41] This technique is applied on recontoured sites to break up compacted layers, however, it may

Figure 9 A ripper at a coal mine.

also be used on a slope to roughen the surface and destroy any slippage plains that might develop between overburden and subsoil or between subsoil and surficial soil layers.

Gouging and dozer basins are terms applied to atypical techniques of seedbed preparation used in arid to semiarid regions. Gouging is the practice of digging a series of small depressions across the landscape (Figure 10). Dozer basins are simply very large gouged depressions. The purpose of both of these depressions is to trap precipitation and snow on the disturbance and to concentrate infiltration in these limited areas. Seedlings have an increased opportunity to survive in dry regions with the augmented water supply in the depressions (Figure 11).

In western mining areas gouged depressions (Figure 12) are usually 10 to 20 cm (4 to 8 in.) deep, 25 to 40 cm (10 to 16 in.) across, and up to 90 cm (35 in.) long.[42] A modification of this technique is used in Australia and other semiarid to arid regions of the world. In these regions a long gouged depression is more common. These depressions may be over 1 m (3.3 ft.) long. They serve the same purposes as the smaller depressions but are oblong rather than oval.

Dozer basins are much larger than gouged depressions. They are made with some sort of attachment to a bulldozer (Figure 13). They may be up to 1 m (3.3 ft) deep, 2.5 m (8 ft) wide, and up to 6 m (20 ft) long.[43] Schematic representations of gouged depressions and dozer basins are shown in Figure 12.

Figure 10 Gouged depressions on a minesoil.

Figure 11 Germinating vegetation on a gouged minesoil.

Gouging and dozer basin construction are carried out after topsoiling and before seeding. Because the resultant soil surface is very rough, drill seeding is not possible so these sites must be broadcast seeded. These sites may be so rough as to require aerial broadcasting or hydroseeding. Despite the apparent rough surface after these treatments, sites manipulated in either manner eventually weather to a smoother topography. Gouged areas are relatively smooth after 5 to

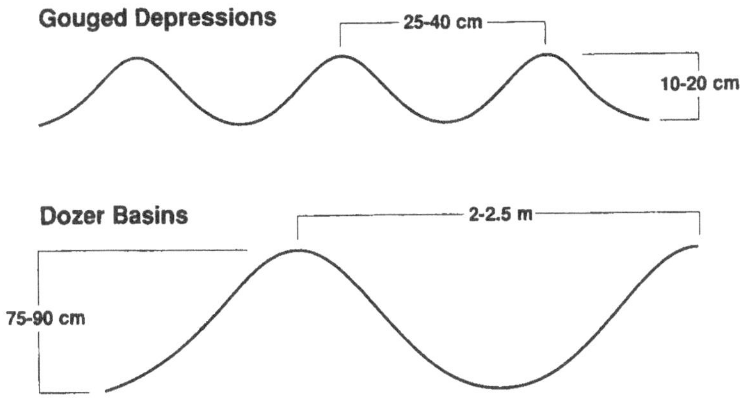

Figure 12 Cross sections of gouged depressions and dozer basins.

Figure 13 Bulldozer attachment for creating catchment basins on a minesoil.

10 years, but dozer basins may remain quite irregular even after 10 years.

Both gouging and dozer basins are usually successful. They decrease erosion and increase infiltration (Table 14). Dozer basins significantly reduced runoff compared to other treatments described in this study.

The influence of the chisel, gouged depressions, and dozer basins on water infiltration are graphically represented in Figure 14. The land on which these plots were located sloped from left to right. Data

TABLE 14 Percent of Precipitation Lost as Runoff at Five Northern Great Plains Mine Sites[44]

	Sites					
Treatment	1	2	3	4	5	Average
Topsoiled						
Chiselled	5.8	18.4	1.2	9.8	3.9	7.8[a]
Gouged	9.5	14.5	1.2	5.0	1.4	6.4[a]
Dozer basins	3.5	6.9	0.9	0.4	1.4	2.4[b]

[a,b] Means followed by different letters are significantly different (p <10%) by paired t test.

collected under the precipitation regimes shown in this figure indicate that the movement of water downslope was faster when the surface was chisel plowed than under the other treatments. Both gouging and dozer basins were better than the chisel plow, concentrating water in restricted areas where it will be available for future plant root absorption. This was the primary goal of these treatments.

Plant establishment and development during the first growing season are concentrated in and adjacent to the depressions (Figure 15). Within two to three growing seasons, bare ground between the gouges or dozer basins develops a vegetation cover from rhizomatous species or seeds produced by the originally seeded plants. Figure 16 illustrates the third growing season on a gouged minesoil in the northern Great Plains. If moisture is limiting vegetation establishment on a disturbance, gouging or dozer basins should be considered viable alternatives to standard seedbed preparation.

AMENDMENTS

Any additions to surface soils or overburden prior to or shortly after dissemination of seed is considered an amendment. In general, amendments provide a better medium for plant growth by causing changes in the rooting medium. These changes include:

1. Macro- or micronutrient additions
2. Organic matter increases
3. Raising or lowering pH
4. Improved hydrologic properties
5. Reduced erosion
6. Temperature modification

Amendments to stimulate or cause these changes may be of several types. Fertilizers to improve the nutrient status of the soil may be composed of inorganic chemicals or organic materials. Organic matter

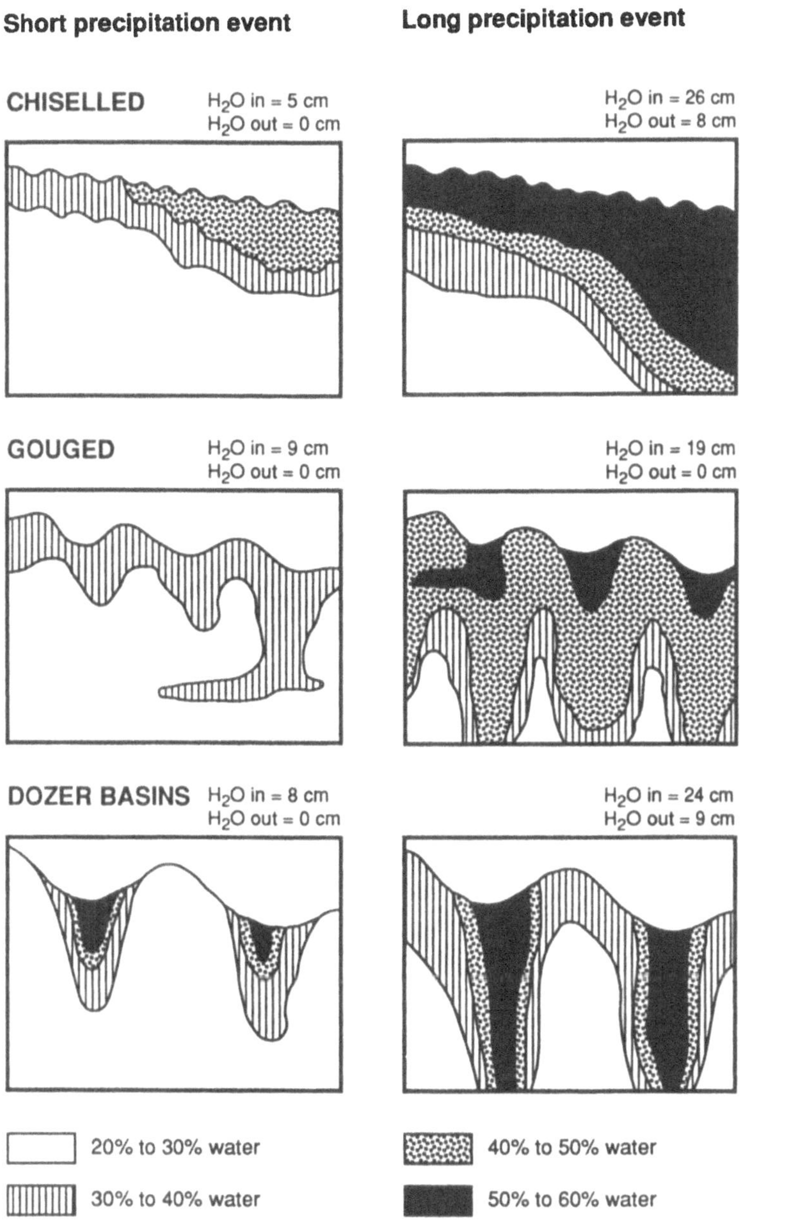

Figure 14 Water-saturated minesoil zones. (Modified from Dollhopf, D. J., Hedberg, D. W., Young, S. A., Goering, J. D., Schafer, W. M., and Levine, C. J., Agric. Exp. Sta. Spec. Rep. 18, Montana State University, Bozeman, 1985, 68.)

Figure 15 First growing season vegetation on gouged minesoil.

Figure 16 Plant response after three growing seasons on a gouged minesoil.

may be made up of materials as diverse as wood chips, straw, paper, sewage sludge, composted municipal waste, or manure. Each of these organic amendments also improves hydrologic properties, reduces erosion, and modifies temperature extremes. Lime or acids may be used to change the pH of the rootzone material. Salts of different types may be added to the soil to encourage the leaching of other undesirable anions or cations.

Fertilizers

Nutrient deficiencies are one of the most universal attributes of degraded soils. While nitrogen deficiency is most common, inadequate supplies of available phosphorus have often been reported in all types of disturbed soils. Potassium deficiencies are not anticipated in young soils but are common in more mature and weathered materials. Micronutrient deficiencies as well as excesses have been found in limited geologic areas of various regions.

Nutrients are found in the soil in different pools. Part of the total soil nutrient pool is readily available to plant roots. This available pool is found in the soil solution. A second portion of the total nutrient content of a soil is complexed in one form or another and slowly becomes available to plants through changes that release the nutrients to the soil solution. These elements may be in organic form or on the cation exchange. The third, and usually largest soil nutrient pool, is insoluble and not available to plants. Only as soil minerals weather does this insoluble part of the soil nutrient supply move into one of the other pools. Fertilizers are usually added to soils as supplements either to the available nutrient pool or to that pool of nutrients that slowly enters the soil solution. Even slow-release fertilizers are considered in these more readily available nutrient reservoirs.

In modern agriculture large amounts of nitrogen, phosphorus, and potassium are added to soils to increase plant growth rates or enhance production of a crop. Figure 17 illustrates the generalized response of a plant to any nutrient. The exact numeric limits for a specific element vary but the shape of this curve is constant for any specific plant.

In land revegetation programs on the Great Plains or the forests of the East, the rehabilitation specialist is most concerned about self-perpetuating plant communities and soil erosion control and less concerned about plant production. The biomass of vegetation produced on a revegetated site is of primary importance only on crop lands. Many naturalized plant species are more productive than native species, but emphasis is placed on species endemic to the disturbed site since they presumably are better adapted to the local climate and soils. When native species are seeded in topsoiled disturbances in the

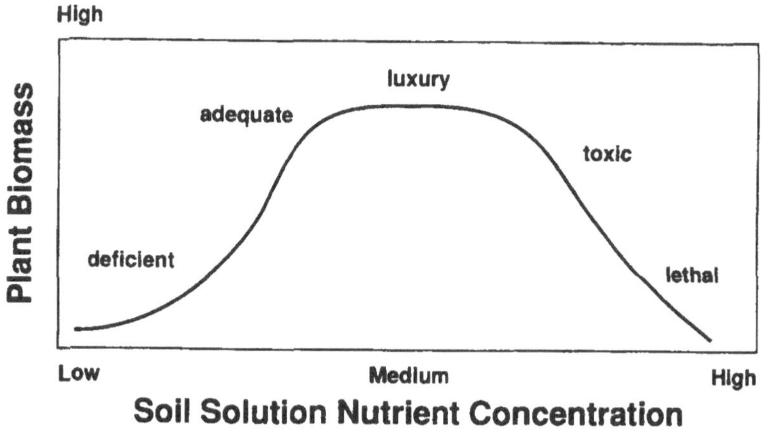

Figure 17 Plant response to nutrient concentrations.

Great Plains, fertilizers are rarely needed. In more mesic regions, and when agronomic species (pasture grasses and legumes) are seeded, fertilization may be necessary to ensure adequate growth.

Nitrogen Fertilization

The use of fertilizers is widespread in disturbed land rehabilitation and the most commonly applied plant nutrient on disturbed lands is nitrogen. This element may be added to the disturbed soil in any of the forms shown in Table 15. Nitrate is the form of nitrogen most readily available to plants. Ammonium ion is also found in very low concentrations in most soils. It is directly absorbed by many plant species, but it is also oxidized to the nitrate in the soil. Nitrogen application rates should be adjusted to meet the requirements of the species being seeded and the particular soil being treated. Rates depend on organic matter content as well as the amount of NO_3-N in the soil.

Crop fertilizer requirements should not be referenced for nitrogen application rates for native grassland plantings. Native perennial vegetation evolved in environments of low nutrient status; therefore,

TABLE 15 Common Fertilizer Forms of N

Name	Chemical Formula	%N
Urea	NH_2CONH_2	46
Ammonium nitrate	NH_4NO_3	35
Anhydrous ammonia	NH_3	82
Ammonium phosphate	$NH_4H_2PO_4$	12 (27% P)
Diammonium phosphate	$(NH_4)_2HPO_4$	21 (23% P)
Ammonium sulfate	$(NH_4)_2SO_4$	21 (24% S)

these species usually have much lower nitrogen requirements than agronomic species. Tame species have been selected, very often, for rapid growth and maximum production and subsequently reveal higher nutrient requirements.

Nitrogen fertilization of disturbed soils is not necessary when the organic matter contents of the surface soil are adequate. Over 99% of the soil nitrogen supply is held in the organic fraction.[46] This reservoir is constantly releasing nitrogen to the soil solution as organic molecules are broken into smaller components by microorganisms. Researchers in Colorado never recommend the addition of nitrogen to native grasslands when the soil organic matter content exceeds 2%.[47] When the organic pool in a soil was less than 1% and soil nitrates were less than 5 μg/g, these authors only recommend 45 kg/ha (40 lb/acre) of nitrogen fertilizer. Research by several investigators supports this very small or no nitrogen recommendation for grasslands seeded on topsoiled minesoils in western coal fields.[48,49] In a recent study in southeastern Montana, the main responses to nitrogen fertilization were increased annual grass or annual weed production and decreased diversity.[50] The nitrogen levels shown in Table 16 are an attempt to place a number on soil nitrogen analyses. These concentrations are only guidelines, however, because postdisturbance land use, organic matter contents, and proposed vegetation strongly influence decisions to amend a site with nitrogen.

Nitrate salts are soluble. When they are used in a fertilizer program, care must be exercised to prevent contamination of ground or surface waters. Because these salts are soluble, when they are applied to the soil surface they are carried to the rootzone by infiltrating precipitation.

A normally functioning nitrogen cycle is essential to successful, long-term revegetation. All components of the nitrogen cycle shown in schematic form in Figure 18 are necessary to the proper functioning cycle. Invertebrates, protozoans, fungi, and bacteria are all active in the decomposition of organic material. When topsoil is stored, organic matter concentrations decrease. Because their energy source is depleted, the number of organisms available to convert organic matter

TABLE 16 Generalized Grassland Mineral N Levels[51]

Rank	kg N/ha[a]
Low	<20
Medium	20–70
High	>70

[a] In the upper 36 cm (15 in.) soil layer (kg/ha × 0.9 = lb/acre).

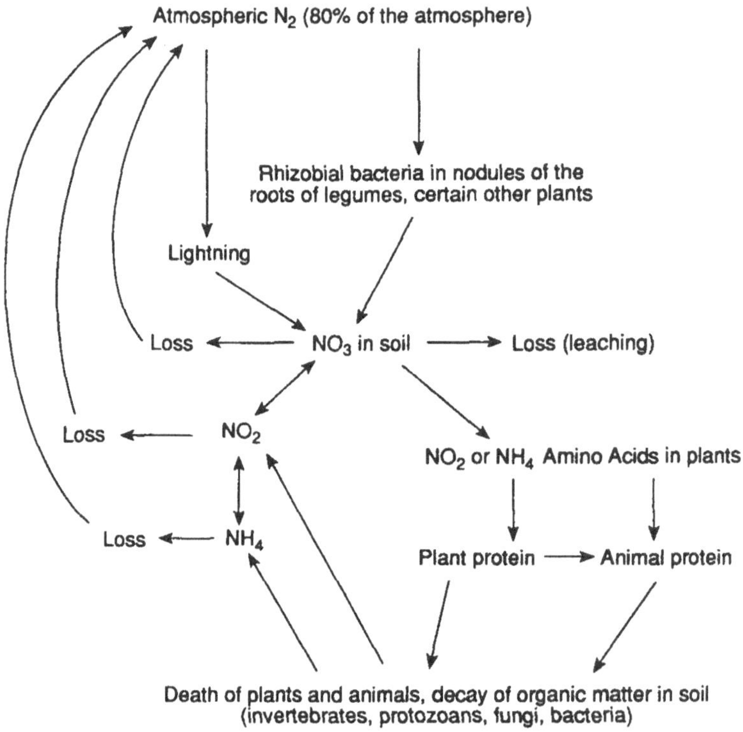

Figure 18 Stylized nitrogen cycle.

to its constituent components also decreases. Addition of organic matter and nitrogen stimulates the development of populations of these macro- and microorganisms to predisturbance levels. The growth of nitrogen-fixing legumes on the disturbed soil also helps to increase nitrogen soil levels and activate the nitrogen cycle. For a more thorough discussion of the role of nitrogen in minesoils, the reader may consult a number of appropriate texts.[52,53]

Phosphorus Fertilization

Like nitrogen fertilization, phosphorus amendment should be based on soil nutrient analyses (Table 17). However, the major soil phosphate reserve is in the organic fraction and is not accessed by chemical extractants. Good quality topsoil will meet the phosphorus as well as the nitrogen needs of most native plant species.

Phosphorus deficiency symptoms are often observed on vegetation growing on disturbed soils from more mesic regions, on hard rock wastes, or on acid soils. In these types of disturbances phosphorus fertilization is very beneficial. Phosphorus is complexed by heavy

TABLE 17 Soil P Levels[54,55]

	P (μg/g soil)	
Rank	Bray ext.	Olsen ext.
Very low	<5	<4
Low	6–14	5–8
Medium	15–25	9–14
High	26–30	15–20
Very high	>30	>20

metals in acid soils. This reaction removes phosphorus from the available soil nutrient pool, but it also removes the heavy metals from this pool. It is highly probable that phosphorus amendment would benefit most soil disturbances of these types.

Phosphate fertilization of semiarid coal minesoils commonly increases the production of pioneer species such as annual grasses and weedy forbs, but produces little or no response by the more desirable perennial grasses.[56] When topsoil is applied over nontoxic subsoils on coal minesoils, phosphorus addition reportedly produces little growth enhancement and can adversely affect diversity.[57]

Legumes used in revegetation programs are nitrogen fixers that have a biological advantage over grasses on nitrogen deficient sites, but these plants also have high phosphorus requirements. If the post-mine land use is legume hay as opposed to the growth of native grasses, phosphorus fertilization will prove beneficial. When the post-mine land use is crop production, fertilization with nitrogen and phosphorus may be necessary to reach maximum production levels. Phosphorus may be added to the soil in any of the forms shown in Table 18.

Phosphorus is less soluble than nitrates. It must be incorporated into the soil with plow, disk, or chisel to ensure good soil-root-phosphorus contact. Therefore, on disturbed sites intended for mixed grass-legume hay production, phosphorus should be incorporated into the soil before seeding.

Potassium Fertilization

This element is rarely deficient in disturbed soils of semiarid regions. Young soils generally are adequately supplied with potassium.

TABLE 18 Common Fertilizer Forms of P

Name	Chemical Formula	%P_2O_5	%P
Superphosphate	$Ca(H_2PO_4)_2$	20	9
Triple superphosphate	$Ca(H_2PO_4)_2$	46	20
Phosphoric acid	H_3PO_4	73	32

Concentrations in Great Plains soils are in the medium to high range (Table 19). Potassium fertilization of coal minesoils or native soils of the northern Great Plains has produced no measurable effect in numerous experiments.[59–61] It is rare, therefore, to find potassium used in fertilizer programs on disturbed sites in the western plains. At disturbances in the eastern coal fields,[62] in the western mountains,[63] as well as in the rehabilitation of mill or smelter wastes,[64] potassium amendment is often necessary. Any of the chemical forms shown in Table 20 may be utilized in these programs.

Fertilizer Mechanics

Nitrogen, phosphorus, and potassium are the major fertilizers. Some sulfur is used in agricultural regions and several micronutrients are added as specific amendments in some countries, but in North America nitrogen, phosphorus, and potassium comprise well over 90% of the fertilizer applied to soils.

By convention, the percent composition of these three minerals is printed on the outside of the fertilizer container. Nitrogen is represented as percent nitrogen, phosphorus as percent phosphorus pentoxide (P_2O_5), and potassium as percent potassium oxide (K_2O). Thus, 20–5–35 printed on a sack of fertilizer indicates that 20% of the contents is nitrogen, 5% is phosphorus pentoxide, and 35% is potassium oxide. The ratio of phosphorus in phosphorus pentoxide is 0.44 and the ratio of potassium in potassium oxide is 0.83. To convert phosphorus pentoxide to elemental phosphorus multiply percent (or weight) phosphorus pentoxide by 0.44, to convert potassium oxide to elemental potassium multiply potassium oxide by 0.83–100 kg of the material

TABLE 19 Soil K Levels (Ammonium Acetate Extraction)[58]

Rank	K Extracted From Soil (μg/g)
Very low	<75
Low	76–125
Medium	126–250
High	251–500
Very high	>500

TABLE 20 Common Fertilizer Forms of K

Name	Chemical Formula	%K_2O	%K
Potassium nitrate	KNO_3	46	39 (14% N)
Potassium sulfate	K_2SO_4	54	45 (18% S)
Potassium chloride	KCl	62	52

in the example shown above (20–5-35) would contain 20 kg of nitrogen, 2.2 kg of phosphorus, and 29 kg of potassium.

The printing of the percentage of nitrogen, phosphorus pentoxide, and potassium oxide on containers of fertilizer is slowly changing in this country and has already changed in most foreign countries. It has been or is being replaced by representation of elemental percentages. Thus, at some time in the future 20–5–35 printed on a container of fertilizer will represent 20% nitrogen, 5% phosphorus, and 35% potassium; but for the present, the label on the container must be consulted to ascertain what is meant by the printed percentages.

Mulches and Organic Amendments

Mulches are organic materials applied to the surface of a disturbance after seeding. The more common mulches include paper, wood residues, straw, and native hay. Manure, compost, and sewage sludge are technically not mulches but rather organic amendments. They are applied to a site prior to seeding and incorporated into the soil. Mulches, which are placed on soils primarily to reduce erosion, have only a limited impact on soil structure and the reestablishment of soil-plant nutrient cycles. Surface mulches reduce wind velocities at the soil surface, shield the soil from raindrop impact, reduce evaporation from the soil surface, trap small soil particles on the site, reduce surface soil temperatures, and help prevent soil crusting.[65] Mulches may be plowed, chiseled, crimped, or rototilled into the soil. When incorporated, mulches act like organic amendments and their impact on soil structure, infiltration, cation exchange capacity, total organic matter, and nutrient cycles is more pronounced.

The application of mulches after seeding may be carried out by a number of methods. These include mechanical dry blowers, modified manure spreaders (Figure 19), and hydromulchers (Figure 20). Newer dry blowers dispense large quantities of mulch. Up to 18 mt/h (20 ton/h) may be dispersed from truck-mounted mulch blowing equipment. This specialized equipment is used to rapidly cover very large areas (several hundred hectares). The use of these powerful dry blowers on most disturbances is not economical because of the expense of the equipment. Manure spreaders replace dry blowers on small parcels of land. When areas of intermediate size (e.g., mines and highway shoulders) are to be treated, hydromulchers are the most economical method of distributing mulch. Hydromulchers mix mulch with a fluid and permit the operator to spray the mixture over the site. Areas that are too small for the large and expensive dry blowers but too large for manure spreaders may be treated very quickly by hydromulchers.

All surface mulches are subject to loss by wind or by overland flow of water. To combat this loss, they are usually held in place by crimp-

Figure 19 Manure spreader.

Figure 20 Hydromulcher.

ing, tackifiers, or netting. Crimping utilizes some sort of mechanical disk, wheel, or punch to push part of the mulch into the seedbed (Figure 21). This is the cheapest and easiest method to hold a mulch in place. It is commonly used after mulching with straw or native hay. The liquid sprayed by the hydromulch machine (Figure 22) is

Figure 21 Crimped straw.

Figure 22 Hydromulch spray.

usually a combination of water, the mulch, and some type of organic polymer or glue-like substance called a tackifier. As it dries, this tackifier binds the mulch particles together and to the surface of the soil. Dyes are also added to hydromulch to enable the operator to estimate application rates. The ability of the hydromulch to disperse a mulch, water carrier, dye, and a binding agent (tackifier) simultaneously contributes to the economy of operation of these machines in large-scale operations. They have been successfully used on slopes as steep as 3:1 or 2:1. Netting is also used to hold a mulch in place, but it is very labor intensive and only used in critical erosion-prone areas. Nets have been successfully used in alpine areas where wind erosion potential is high[66] and on very steep slopes (2:1 or steeper).[67]

As a general rule, and within certain limits, long particles of mulch are most beneficial and their benefits last longer than shorter particles.[68,69] Unfortunately, the fast methods of mulch application, dry blowers and hydromulchers, require small particles. When these types of equipment are used for mulch application, a compromise must be made between particle size and benefits of the application. New tackifiers and hydromulches are more efficient and effective, so the advantages of long particles are becoming less obvious as manufacturers improve these products.

Almost all mulches are light and bulky. While the source (e.g., woodchips and straw) may be inexpensive, large quantities are needed and transportation costs may prohibit their use in remote areas. Therefore, it may be cheaper to transport shredding devices to isolated areas to mulch organic products available on site rather than to haul organic matter to the disturbance.

Sewage sludge, compost, manure, and other organic amendments provide a ready source of carbon and usually nitrogen for microorganisms in the soil. Populations of bacteria, fungi, actinomycetes, and protozoans involved in the decomposition of organic material increase dramatically when such materials are added to soils. These microorganisms break down the organic matter and provide a supply of nutrients for higher plants. However, when nitrogen-poor organic amendments are added to a soil, microorganism numbers increase as much as the energy supply permits, but they consume a major portion of the available soil nitrogen. This element may become limiting to microorganisms and to higher plant growth. When the C:N ratio exceeds 20:1, grassland vegetation begins to reveal the effects of nitrogen deficiency. When the C:N ratio approaches and exceeds 30:1, microorganisms themselves experience a nitrogen deficit.

Disruption of the C:N ratio by the addition of large amounts of organic material is relatively common on degraded lands. The C:N ratio of several substances commonly added to disturbed soils are shown in Table 21. Nitrogen must be added to soils when organic amendments such as straw or wood residues are used. As a general rule, nitrate additions should be sufficient to maintain a C:N ratio between 12:1 and 20:1. Thus, the addition of 1 mt/ha of sawdust would necessitate the fertilization of the soil with 50 kg/ha N, or more, to maintain a C:N ratio within this range.

Regulatory agencies often require the application of organic matter or mulches to disturbed lands. This material is said to improve soil

TABLE 21 C:N Ratios for Some Common Materials[70]

Material	Nitrogen (%)	C:N Ratio
Raw sewage sludge	4–7	11
Grass clippings	3–6	12–15
Cow manure	1.7	18
Horse manure	2.3	25
Oat straw	1.1	48
Wheat straw	0.3–0.5	130–150
Sawdust	0.1	200–500

structure, increase infiltration, reduce erosion, lower soil surface temperatures, as well as encourage the germination of seeds and establishment of seedlings. Many disturbances (e.g., sand and gravel pits, highway rights-of-way, and mine wastes) need additional organic matter in the surface soil layer. Nevertheless, amendment of reconstructed soils with organic matter does not produce benefits compatible with its cost if the degraded site is covered with topsoil that has an adequate organic matter pool and is located on relatively flat ground. That is, good quality, direct-hauled topsoil, on a 5:1 slope usually does not need supplemental organic matter. The addition of organic materials to such a site only necessitates additional costs in the form of fertilizers to correct the increased C:N ratio produced by the organic matter. While organic amendments are beneficial on many materials, regulations requiring their addition to all disturbances increase reclamation costs without a concurrent increase in the quality of the reclamation program.

Paper

This mulch, which is usually composed of recycled paper, is the cheapest of the organic amendments. Fibers are usually very short. It is commonly applied with the hydromulcher at rates of 1700 kg/ha (1500 lb/acre) or more. Even when applied with a tackifier, however, particles tend to detach and float away on runoff water. Paper is not as effective at reducing surface temperature or soil moisture loss as wood residues.[71] It contains no nitrogen and must be supplemented with nitrogen fertilizer. Decomposition rates are extremely fast and surface benefits of this mulch are short lived. Paper should not be used where surface runoff is a problem.

Wood Residues

These include wood chips or fibers, sawdust, and bark fragments. These materials are a cheap organic material for addition to soils. They are applied at rates of 1700 kg/ha (1500 lb/acre) or more, usually with a tackifier, or they are incorporated like other organic amendments. As a mulch, they increase seedling establishment and reduce soil erosion on slopes up to 3:1.[72] On steeper slopes, they often produce unacceptable results. Wood products contain little or no nitrogen and upset the C:N ratio when incorporated into the soil. Nitrogen fertilization is obligatory with this type of amendment.

The benefits of sawdust are reduced by the rapid digestion of this material by microbes and its undesirable C:N ratio. Chips and bark fragments last much longer and are the more common form of wood residue used as an organic amendment. They may be applied with any of the mechanical mulching devices. Fibers are often placed on the soil surface as a mulch or they may also be applied as some type of woven erosion control mat. If applied by dry blowers or as hydromulch, wood particles must be ground to an appropriate size, thereby increasing the cost of application and decreasing their effectiveness. Wood particles may be stabilized with some type of tackifier or they may be plowed into the soil.

Wood residues (chips, sawdust, and bark) at rates of 67 mt/ha (30 ton/acre), and with supplemental nitrogen, were widely used in the rehabilitation of abandoned bentonite mine lands in the northern Great Plains. The effectiveness of these amendments and the rates at which they should be applied were demonstrated in research in Wyoming.[73]

Straw

This material is primarily composed of the stems of cereal grains such as wheat *(Triticum aestivum)* or oats *(Avena sativa)*. Application rates for loose straw are 1.1 to 4.5 mt/ha (0.5 to 2 ton/acre).[74–76] The rate should be adjusted for slope, erosion potential, and other site-specific characteristics. Like wood residues, straw contains almost no nitrogen, so fertilization is necessary with this amendment. Straw decomposes even more rapidly than wood products, therefore it is imperative that the stems be as long as possible to prolong the life of the mulch. Straw usually contains many seeds of the parent crop. When a straw mulch is applied to a site, a nurse crop of the mulch species is inadvertently seeded. Fortunately, in cooler climates the cover crop persists for only one growing season. Of greater concern are the seeds of weeds that are also brought to the disturbed site with the straw. Straw must be selected with the greatest care to reduce the amount of weed seed introduced with this mulch. The use of straw has been recommended or required by many state agencies responsible for disturbed land revegetation.

Some action must be taken to hold straw on site after distribution. Straw stabilization is usually accomplished with a crimper. This machine pushes part of the straw 5 to 7.5 cm (2 to 3 in.) into the ground (Figure 23). The protruding stem (Figure 21) acts like stubble: it shades the ground, breaks the force of wind or rain, and funnels precipitation into the soil. Tackifiers and netting have also been used with straw mulches. Straw held on site by any of the three methods provides an excellent means of reducing surface wind velocity, increasing infiltration rates, and decreasing surface runoff and soil temperatures.

Native Hay

This material is very similar to straw, reveals similar disadvantages, and is applied at identical rates, but it has several distinct advantages.[77] If properly cut and dried, native hay stems are longer and stronger than grain straw. It also contains seeds of native plants rather than cereal grains. These seeds are beneficial to the development of

Figure 23 Crimping straw.

a diverse plant community on the disturbance. Because this mulch contains many leaves, flowers, and seed heads, it often contains a small amount of nitrogen. Because of this residual nitrogen it requires no fertilizer amendment, or less fertilizer than most other mulches. Native hay must be distributed with a modified manure spreader or specialized hay spreader. If ground to pass through the dry blower or hydromulcher, it loses most of its desirable properties.

The collection of native hay must be carried out with care. Seeds of undesirable plants are more readily transmitted to the disturbed site in native hay than in straw. Small, undesirable plants such as the annual brome grasses *(Bromus tectorum* and *B. japonicus)* may be disseminated in native hay mulches.

The use of native hay on coal mined lands in the northern Great Plains was studied in 1981–1982. The study revealed that seedlings of native species for which there was no commercial seed source could be obtained by harvesting and utilizing native hay mulches.[78] More diverse communities can be developed on disturbances mulched with native hay than on sites mulched with straw, wood residues, or paper products.

Stubble Mulch

This term refers to the stems and roots remaining after growth of a cereal grain or an annual crop on a disturbed site. These plants are often used as an organic amendment and nurse crop. Because they are seeded and grow on the site, they are less likely to be lost to wind erosion and last longer than crimped straw mulch, but they provide the plant growth enhancing attributes of crimped straw.[79] Stubble mulches are usually seeded in the spring and the permanent vegetation cover interseeded during the fall of the same growing season. The types of drills necessary to get seed into the ground in this stubble limit the number of species that can be seeded in stubble mulches.

A modification of this technique utilizes a cereal grain or other annual plant species as a nurse crop. The annual species is seeded with the permanent vegetation and acts much like a stubble mulch, but only one seeding is required. When cereal grains are used in this capacity they must be seeded at very light rates (i.e., 10 kg/ha [9 lb/acre]).

Erosion Control Blankets

These devices, also called mats, are a type of surface mulch. Mats are constructed of an organic material such as straw (Figure 24), wood fibers (excelsior), coconut fiber (Figure 24), jute (Figure 25), or burlap woven together or held in place by some woven material. They are designed to control erosion on slopes.[80] Vegetation may be seeded

Figure 24 Two examples of erosion control blankets. Top: Straw mat. Bottom: Coconut fiber mat.

Figure 25 Woven jute erosion control blanket.

under the mat or seed may be impregnated into the organic material. In either case, erosion control blankets are designed so that seeds germinate and stems grow through and above the mat. As the fabric ages it becomes incorporated into the soil, decomposes, and loses its ability to control sediment loss from the slope. However, it is replaced by living plants that continue the control of erosion initiated by the mat.

Common forms of erosion control blankets may consist of a woven jute mat or a core of straw surrounded by two layers of plastic netting (Figure 24). Variable amounts of straw or different types of biological materials (i.e., coconut fiber or burlap) may be used for different slope angles, overland flow rates, soil materials, erosion potential, or post-disturbance land use. Mats require close soil contact and are usually attached to the seed bed with large staples. The edges of the mat must be buried in the soil to prevent wind from lifting the blanket from the soil surface. If wind gets under the fabric it destroys soil-mat contact and the effectiveness of the erosion control blanket.

Mats are expensive but effective. Their costs may be many thousands of dollars per hectare, but erosion reduction could represent up to 90% of soil loss on unprotected soils.[81] The cost of mats prohibits widespread use on most disturbed sites. They have, nevertheless, been successfully used on steep slopes and erosion-prone soils throughout the world and they may be the only realistic alternative on many slopes.

Manure

Livestock waste provides a valuable source of soil organic matter and nutrients. Where available in adequate quantities, it is one of the more complete organic additives that can be applied to a disturbed soil.[82] For best results, it should be plowed into the soil after distribution with a manure spreader (Figure 19). Part of the manure readily decomposes and provides both carbon and nitrogen to soil microorganisms and nitrogen to higher plants. Animal wastes do not upset the nitrogen balance unless they contain an exceptional amount of straw. Manure is especially useful in acid wastes. It apparently complexes much of the toxic metal load in such soils and removes these metals from the available soil elemental pool.

If salinity is a problem in the disturbed soil, manure additions should be carefully assessed since animal wastes possess a high salt load. Addition of this material to a saline site exacerbates the salt problem.

Green Manure

The growth of plants and their subsequent plowing into the soil is referred to as a green manure crop. This technique is used to increase soil organic matter while stabilizing surface soils and inhibiting development of weedy species. In most uses, the crop is permitted to develop for one or more growing seasons before it is plowed under and the perennial vegetative cover is seeded. Legumes or small grains are most often used for green manure crops, but any plant that produces an abundant root system and a large amount of above ground biomass in a short time (one or two growing seasons) is an adequate candidate for such a crop. Legumes are favored for this purpose when soils are nutrient deficient because these plants are nitrogen fixers and contribute nitrogen and nitrogen-rich organic matter to the soil. This technique is used extensively on agricultural lands and has been successfully applied on minesoils and other disturbances throughout the world.

Compost

This material is derived from the biological decomposition of any solid waste. Decomposition may be brief or may have lasted several months, but the compost should be relatively stable to avoid undesirable odors. That is, the initial, rapid breakdown of the waste by microorganisms should be complete. Changes now taking place in the compost should be proceeding at a relatively slow rate. The original source of the organic matter may be anything from grass clippings, to municipal solid waste, to sewage sludge, but in each example the compost that results is an excellent soil conditioner. Compost has been found to enhance macronutrient fertility, improve soil structure, increase the infiltration of precipitation and the water holding capacity, decrease bulk density, and increase the cation exchange capacity of the receiving soil, subsoil, or overburden.[83]

Compost, as a low-cost soil fertilizer and conditioner, has received widespread use on degraded sites and agricultural lands throughout the world. It has been used for centuries in Asia and Europe on disturbed soils and agricultural lands. Its use on similar sites in North America is relatively new and has been limited. Recent research in the Northwest clearly reveals the benefits of compost additions to arid rangelands.[84] The authors of this research suggest that the remarkable benefits of compost in these revegetation trials may be related to its ability to augment the water-holding capacity of the study site soils.

When a surface soil is not available for use in a rehabilitation program, compost is especially useful as an amendment to enhance soil structure and nutrient composition of the substitute materials. As a general rule, to impact soil structure application rates should be approximately 25% by volume.[85]

Carbon:nitrogen ratios are good indicators of compost fertility. The ratios are almost always less than 20:1, but they vary with the contents of the original composted material. If this ratio is low enough (<10 or 12:1) compost may be used as a fertilizer as well as a soil conditioner.

Almost all municipalities in this country are developing or have established composting centers as alternatives to landfilling organic wastes. Quantities of compost are rapidly increasing and the application of this material to degraded lands will be limited less by expense and transportation costs as the composting of organic wastes expands in the future.

Sewage Sludge

This is the material (organic and inorganic) removed from waste water at sewage treatment plants. Like compost, it is available in growing quantities near most urban centers. Although sewage sludge begins as a material made up primarily of water, its solids content can be increased by any of several successive steps. These range from settling solids in the primary treatment of sewage to heating aerobically digested wastes. This sludge contains a number of nutrients in addition to organic residues. The unique combination of nutrients and organic matter make sewage sludge a valuable addition to the amendments for degraded sites.[86] The effect of sewage sludge on soils is similar to that of compost: it improves soil structure, increases the water holding and cation exchange capacities, reduces bulk density and surface temperatures, ameliorates pH extremes, and contributes nutrients to the receiving soil, subsoil, or spoil.

Topsoil may be amended with sewage sludge or built up from subsoils by the addition of sludges. These wastes may be applied as a liquid or as a solid. Liquids are allowed to percolate into the soil or are injected into the rootzone. When applied as a solid the sludge must be mixed into the soil, usually with a plow, disk, or chisel. Although the most common method of sludge utilization requires mixing the waste with soil, studies in Pennsylvania have shown very good results by using a mixture of sewage sludge and fly ash as a topsoil or rooting medium on severely denuded and eroded slopes.[87,88]

The nutrient analysis of sewage sludges varies widely but a nitrogen:phosphorus:potassium ratio of 5.0:2.5:0.4 has been suggested as typical of many sludges.[89] This is a relatively well-balanced grassland fertilizer. Unfortunately, several potentially phytotoxic metals are often in elevated concentrations in these materials. These metals include copper, cadmium, chromium, nickel, lead, and zinc.[90,91] Continued application of sewage sludge to a site may produce elevated levels of one or more of these metals in the soil in addition to the increased macronutrient levels.

Metals in sewage sludge pose the main limiting factor to the use of this material, or compost made from sludge, as soil amendments. The federal government has recommended maximum amounts of metals that should be deposited in or on agricultural soils based on the cation exchange capacity of the receiving soil (Table 22). These loading rates are based on the possible accumulation of elevated metal levels in plants and animals in food chains leading to man. Sludge loading rates (kg/ha × 0.9 = lb/acre) may be calculated from the formula:

$$\text{Sludge Loading Rate} = \frac{\text{RMMLR (mg/ha)}}{\text{MCS (mg/Kg)}} \qquad [3]$$

in which RMMLR = Recommended Maximum Metal Loading Rate in mg/ha; and
MCS = Metal Concentration in the Sludge in mg/kg.

Studies of food chain transport of these metals have revealed elevated metal levels in soils, crops, and animals ingesting the crops.[93–95] Within the limits established by federal guidelines, however, sludge may be applied to agricultural lands without exceeding tolerable limits for trace element concentrations in the human diet. On barren mined lands or severely polluted soils that will not grow food for human consumption, large amounts of these amendments may be added to the soils with the assurance that humans will not be exposed to elevated trace element levels resulting from that use.

Lime

Lime is a generic term for almost any calcium amendment that ameliorates soil acidity and releases calcium ions into the soil. The most common material called lime is ground limestone, or calcium

TABLE 22 Federal Recommendations For Maximum Trace Element Loadings for Agricultural Lands[92]

	Cation Exchange Capacity (meq/100 g)		
Metal	<5	5 to 15	>15
Cd	5[a]	10	20
Cu	125	250	500
Ni	50	100	200
Pb	500	1000	2000
Zn	250	500	1000

[a] kg/ha (kg/ha × 0.9 = lb/acre)

carbonate ($CaCO_3$). Calcium hydroxide [$Ca(OH)_2$] is also used for or in conjunction with limestone. Pure lime is calcium oxide (CaO). It is only occasionally used as a soil additive.

Amendment with these materials is the common treatment for acid soils produced by agriculture or industry.[96] Soils of humid regions, especially those cropped for many years, become acid because of the loss of bases and an accumulation of hydrogen ions on exchange sites and in the soil solution. Violently disturbed soils may become acid because of the oxidation of acid-generating materials previously buried deep in the soil profile. Iron pyrite is the most notorious acid-generating spoil material. When this compound is in an anaerobic environment it does not generate acid, but when exposed to the atmosphere and water during the extraction of a mineral deposit, it begins to oxidize and produce hydrogen ions. Increasing acidity in the new soil reduces the availability of macronutrients and increases the availability of many metals. Metal levels often reach toxic concentrations in this environment and prevent development of a healthy plant community on the disturbance.

Because lime is relatively insoluble, it must be ground to very small size and tilled into the rootzone to exert its maximum effect on soil acidity (Figure 26). Usually, lime is passed through a 50-mesh or finer screen before application to the acid soil. Larger particles react with acid, but complete reaction may take months or years. Some large lime particles may become coated with precipitates so that reaction is inhibited or prevented altogether.

Liming rates for disturbed lands are sometimes based on soil pH (a measure of hydrogen ion concentration in the soil solution) and sometimes on hydrogen ion saturation of cation exchange sites. Acid soils created by mining disturbances and limed on the basis of soil pH and exchangeable acidity alone often reacidify.[97,98] Oxidation of residual pyrite generates enough acid to neutralize the lime and reduce the pH of the soil to its preliming level. Some sort of periodic amendment program must be established to maintain vegetation on these sites. A measure of total residual acid production potential must be made if the soil is to be successfully amended with only one lime application.

Various methods of estimating residual acid-generating capacity have been developed. Identification of sulfur fractions and calculation of an acid-base account works well in eastern coal fields but has not been equally effective in the West. Weathering chambers in combination with sulfur speciation and the determination of an acid-base balance has met with some success in the more arid western minesoils.

Figure 26 Plowing an amendment (lime) into an acid minesoil.

Sodic Soil Amendments

High sodium soils occur on disturbed sites in semiarid to arid regions. Several amendments have been used to ameliorate sodicity. These include calcium chloride ($CaCl_2$), magnesium chloride ($MgCl_2$), gypsum ($CaSO_4 \cdot 2H_2O$), and phosphogypsum. Magnesium chloride and phosphogypsum are industrial wastes that are not widely used as soil amendments. Calcium chloride is the most expensive of these amendments. It is also more soluble than gypsum. Both calcium chloride and gypsum supply calcium ions to the amended material to lower sodium absorption ratios. Addition of these amendments tends to flocculate the soil, increase infiltration, and permit leaching of salts deeper into the profile. They also reduce surface soil crusting and thereby enhance revegetation. Studies in Wyoming[99] and Montana[100] have revealed the improvement of soils and increased production of vegetation when these inorganic chemicals were added to the seed bed.

IRRIGATION

Additional water is occasionally added to degraded lands after seeding (Figure 27). This water may be supplied by drip, sprinkler, or flood irrigation. Generally up to 4 cm (1.5 in.) are applied in spring or early summer. Irrigation water encourages seed germination, plant establishment, and plant growth.[101] It also extends the planting or seeding season. Like matting, irrigation is expensive, but unlike matting, its long-term benefits are questionable. Researchers have shown that irrigation produces benefits during the first few growing seasons.[102,103] As the number of years between water application and response measurement increases, however, the benefits of supplemental water for vegetation establishment become less clear.

Irrigation should only be applied to disturbances to enhance seed germination and plant establishment when natural rainfall is inadequate. The decision to use artificial water sources must be carefully evaluated. It may be less expensive to reseed a stand and wait for the next wet season than to set up an irrigation system and pump water to the site. Furthermore, prolonged application leads to the development of a plant community dependent on supplemental water for survival. Termination of water inputs into such a system results in drastic changes in community production and composition.

Figure 27 Sprinkler irrigation on mined lands.

REFERENCES

1. Workman, J. P. and Tanaka, J. A., Economic feasibility and management considerations in range revegetation, *J. Range Manage.*, 44, 566, 1991.
2. Thornburg, A. A., Plant Materials for Use on Surface-Mined Lands in Arid and Semiarid Regions, S.C.S. TP-157, U.S. Department of Agriculture, Washington, D.C., 1982, 5.
3. Weaver, J. E., *Root Development in the Grassland Formation,* Publ. No. 292, Carnegie Institute, Washington, D.C., 1920.
4. Coupland, R. T. and Johnson, R. E., Rooting characteristics of native grassland species in Saskatchewan, *J. Ecol.*, 53, 475, 1964.
5. Foxx, T. S., Tierney, G. D., and Williams, J. M., Rooting Depths of Plants Relative to Biological and Environmental Factors, LA-10254-MS, Los Alamos National Laboratory, Los Alamos, NM, 1984.
6. Elliott, J. D., Geomorphic Evaluation of Erosional Stability at Reclaimed Surface Mines in Northwestern Colorado, Water Resour. Invest. Rep. 90–4132, U.S. Geological Survey, Denver, CO, 1990, 7.
7. Elliott, J. D., Water Resour. Invest. Rep. 90–4132, U.S. Geological Survey, Denver, CO, 1990, 18.
8. Elliott, J. D., Water Resour. Invest. Rep. 90–4132, U.S. Geological Survey, Denver, CO, 1990, 20.
9. Norbeck, P. M., Stearns, M. W., and Mellbom, P. M., Assessment of infiltration rates for disturbed and undisturbed sites at the Rosebud Mine, Colstrip, Montana, *Proc. Mont. Acad. Sci.*, 43, 111, 1984.
10. Meyer, L. D., Foster, G. R., and Romkens, M. J., Sources of soil eroded by water from upland slopes, in Present and Prospective Technology for Prediction of Sediment Yields and Sources, Agric. Res. Serv. Rep. ARS-S-40, U.S. Department of Agriculture, Washington, D.C., 1975, 177.
11. Farmer, E. E. and Blue, W. G., Reclamation of lands mined for phosphate, in *Reclamation of Drastically Disturbed Lands,* Schaller, F. W. and Sutton, P., Eds., American Society of Agronomy, Madison, WI, 1978, 585.
12. Montana Coal Bureau, Strip and underground mine reclamation rules and regulations, in Administrative Rules of Montana, 26.4.501a Final Grading Requirements, Montana Department of State Lands, Helena, 1988, 26–518.
13. Spry, M. J., Revegetation Research on Hard Rock Mining Disturbances in North-Central Montana, M.S. thesis, Montana State University, Bozeman, 1986, 43.
14. Elliott, J. D., Geomorphic Evaluation of Erosional Stability at Reclaimed Surface Mines in Northwestern Colorado, Water Resour. Invest. Rep. 90–4132, U.S. Geological Survey, Denver, CO, 1990, 7.
15. Zimpfer, G. L., O'Brien, M., and Schumm, S. A., *Design of Stable Drainage Networks for Mined Land Reclamation,* Final Rep. Col. Comm. Higher Educ., Colorado State University, Fort Collins, 1982, 11.
16. Elliott, J. D., Geomorphic Evaluation of Erosional Stability at Reclaimed Surface Mines in Northwestern Colorado, Water Resour. Invest. Rep. 90–4132, U.S. Geological Survey, Denver, CO, 1990, 38.

17. McIntosh, S., A combination of techniques for reconstruction of ephemeral stream channels in Wyoming, in *Proc. Conf. "Reclamation, A Global Perspective." Vol. I,* Alta. Land Conserv. Reclam. Coun., Rep. No. RRTAC-89–2, Calgary, 1989, 229.
18. Power, J. F., Ries, R. E., and Sandoval, F. M., Use of soil materials on spoils — effects of thickness and quality, *N.D. Farm Res.*, 34(1), 23, 1976.
19. Doll, E. C., Merrill, S. D., and Halvorson, G. A., *Soil Replacement for Reclamation of Stripmined Lands in North Dakota,* Agric. Exp. Sta. Bull. 514, North Dakota State University, Fargo, 1984, 5.
20. Montana Department of State Lands, Suspect Levels of Soil Parameters, Coal and Uranium Bur., Montana Department of State Lands, Helena, 1977, 1.
21. Doll, E. C., Merrill, S. D., and Halvorson, G. A., *Soil Replacement for Reclamation of Stripmined Lands in North Dakota,* Agric. Exp. Sta. Bull. 514, North Dakota State University, Fargo, 1984, 4.
22. Ross, D. J. and Cairns, A., Nitrogen availability and microbial biomass in stockpiled topsoils in Southland, *N.Z. J. Sci.*, 24, 137, 1981.
23. Visser, S., Fujikawa, J., Griffith, C. L., and Parkinson, D., Effect of topsoil storage on microbial activity, primary production, and decomposition potential, *Plant Soil,* 82, 41, 1984.
24. Ross, D. J. and Cairns, A., Nitrogen availability and microbial biomass in stockpiled topsoils in Southland, *N.Z. J. Sci.*, 24, 137, 1981.
25. Elliott, G. L. and Veness, J. A., Some effects of stockpiling topsoil, *J. Soil Conserv.*, Jan. 6, 1985.
26. Elliott, G. L. and Veness, J. A., *J. Soil Conserv.*, Jan. 6, 1985.
27. Dickie, J. B. and Gajjar, K. H., The survival of viable seeds in stored topsoil from opencast coal workings and its implications for site restoration, *Biol. Conserv.*, 43, 257, 1988.
28. Thurber Consultants Ltd., Land Resources Network Ltd., and Norwest Soil Res. Ltd., *Review of the Effects of Storage on Topsoil Quality,* Alta. Land Conserv. Reclam. Coun. Rep. No. RRTAC 90–5, Edmonton, 1990.
29. Adbul-Kareem, A. W. and McRae, S. G., The effects on topsoil of long-term storage in stockpiles, *Plant Soil,* 76, 357, 1984.
30. Williams, E. D., *Rehabilitation of Fire Suppression Impacts on the North Fork Fire in Yellowstone National Park,* paper presented at the Am. Soc. Surf. Min. Reclam. Meet., Drango, CO, 1991.
31. Power, J. F., Sandoval, F. M., and Ries, R. E., Topsoil-subsoil requirements to restore North Dakota mined lands to original productivity, *Min. Eng.*, Dec., 1708, 1979.
32. Barth, R. C. and Martin, B. K., *Soil-Depth Requirements to Reclaim Surface-Mined Areas in the Northern Great Plains,* Res. Inst., Colorado School of Mines, Golden, 1982, 11.
33. Power, J. F., Sandoval, F. M., and Ries, R. E., Topsoil-subsoil requirements to restore North Dakota mined land to original productivity, *Min. Eng.*, Dec., 1708, 1979.
34. Merrill, S. D., Doering, E. J., Powers, J. F., and Sandoval, F. M., Sodium movement in soil/minesoil profiles: diffusion and convection, *Soil Sci.*, 136, 308, 1983.

35. Dollhopf, D. J., Postle, P. C., Rennick, R. B., and Young, S. A., *Chemical Amendment and Irrigation Effects on Sodium Migration and Vegetation Characteristics in 7-Year Old Sodic Minesoils,* Agric. Exp. Sta. Spec. Rep. 17, Montana State University, Bozeman, 1985, 107.
36. Dollhopf, D. J., Olson, B. E., and Jennings, S. R., *Coversoil Sodication and Plant Performance on 11-Year Old Minesoils,* Reclam. Res. Unit Publ. No. 9006, Montana State University, Bozeman, 1992, 4.
37. Wyatt, J. W., Dollhopf, D. J., and Schafer, W. M., Root distribution in 1 to 48-year old stripmine spoils in southeastern Montana, *J. Range Manage.,* 33, 101, 1980.
38. Weaver, J. E., *Root Development in the Grassland Formation,* Publ. No. 292, Carnegie Institute, Washington, D.C., 1920.
39. Coupland, R. T. and Johnson, R. E., Rooting characteristics of native grassland species in Saskatchewan, *J. Ecol.,* 53, 475, 1964.
40. Foxx, T. S., Tierney, G. D., and Williams, J. M., Rooting Depths of Plants Relative to Biological and Environmental Factors, LA-10254-MS, Los Alamos National Laboratory, Los Alamos, NM, 1984.
41. USDA Forest Service, User Guide to Vegetation, For. Serv. Gen. Tech. Rep., Int-64, Interm. For. Range Exp. Sta., U.S. Department of Agriculture, Ogden, UT, 1979, 29.
42. Dollhopf, D. J., Hedberg, D. W., Young, S. A., Goering, J. D., Schafer, W. M., and Levine, C. J., *Effects of Surface Manipulation on Mined Land Reclamation,* Agric. Exp. Sta. Spec. Rep. 18, Montana State University, Bozeman, 1985, 7.
43. Dollhopf, D. J., Hedberg, D. W., Young, S. A., Goering, J. D., Schafer, W. M., and Levine, C. J., *Effects of Surface Manipulation on Mixed Land Reclamation,* Agric. Exp. Sta. Spec. Rep. 18, Montana State University, Bozeman, 1985, 11.
44. Dollhopf, D. J., Hedberg, D. W., Young, S. A., Goering, J. D., Schafer, W. M., and Levine, C. J., *Effects of Surface Manipulation on Mixed Land Reclamation,* Agric. Exp. Sta. Spec. Rep. 18, Montana State University, Bozeman, 1985, 41.
45. Dollhopf, D. J., Hedberg, D. W., Young, S. A., Goering, J. D., Schafer, W. M., and Levine, C. J., *Effects of Surface Manipulation on Mixed Land Reclamation,* Agric. Exp. Sta. Spec. Rep. 18, Montana State University, Bozeman, 1985, 68.
46. Woodmansee, R. G., Vallis, I., and Mott, J. J., Grassland nitrogen, *Ecol. Bull.* (Stockholm), 33, 443, 1981.
47. Soltanpour, P. N., Follett, R. H., and Westfall, D. G., *Guide to Fertilizer Recommendations in Colorado,* Coop. Ext. Serv., Colorado State University, Fort Collins, 1985, 31.
48. DePuit, E. J. and Coenenberg, J. G., *Responses of Revegetated Coal Stripmine Spoils to Variable Fertilization Rates, Longevity of Fertilization Program, and Season of Seeding,* Agric. Exp. Sta. Res. Rep. 150, Montana State University, Bozeman, 1979.
49. Rennick, R. B., Hertzog P. J., and Munshower, F. F., *Native Species Response to Fertilizers on Surface Mined Land,* Agric. Exp. Sta. Spec. Rep. 11, Montana State University, Bozeman, 1984.

50. Hertzog, P. J., Response of Native Species to Variable Nitrogen, Phosphorus, and Potassium Fertilization on Mine Soils, M.S. thesis, Montana State University, Bozeman, 1983.
51. Woodmansee, R. G., Vallis, I., and Mott, J. J., Grassland nitrogen, *Ecol. Bull.* (Stockholm), 33, 443, 1981.
52. Reeder, J. D. and Sabey, B., Nitrogen, in *Reclaiming Mine Soils and Overburden in the Western United States,* Williams, R. D. and Schuman, G. E., Eds., Soil and Water Conservation Society, Ankeny, IA, 1987, 155.
53. Hardarson, G., Danso, S. K., and Zapata, F., Biological nitrogen fixation in field crops, in *CRC Handbook of Plant Science in Agriculture,* Christie, B. R., Ed., CRC Press, Boca Raton, FL, 1987, 165.
54. Evangelow, V. P., Barnhisel, R. I., and Thom, W. O., Laboratory spoil tests and interpretation for assessing fertility levels, *Reclam. News and Views,* Coop. Ext. Serv., University of Kentucky, Lexington, 3, 1, 1985.
55. Energy Laboratories, *Soil Analysis Interpretation Guide,* Billings, MT.
56. Rennick, R. B., Hertzog, P. J., and Munshower, F. F., *Native Species Response to Fertilizers on Surface Mined Land,* Agric. Exp. Sta. Spec. Rep. 11, Montana State University, Bozeman, 1984.
57. Hertzog, P. J., Response of Native Species to Variable Nitrogen, Phosphorus, and Potassium Fertilization on Mine Soils, M.S. thesis, Montana State University, Bozeman, 1983.
58. Energy Laboratories, *Soil Analysis Interpretation Guide,* Billings, MT.
59. Wilson, R. and Wambolt, C., *Native Range Fertilizer Guide,* Folder 224, Coop. Ext. Serv., Montana State University, Bozeman, 1981, 1.
60. Hertzog, P. J., Response of Native Species to Variable Nitrogen, Phosphorus, and Potassium Fertilization on Mine Soils, M.S. thesis, Montana State University, Bozeman, 1983.
61. Rennick, R. B., Hertzog, P. J., and Munshower, F. F., *Native Species Response to Fertilizers on Surface Mined Land,* Agric. Exp. Sta. Spec. Rep. 11, Montana State University, Bozeman, 1984.
62. Evangelow, V. P., Barnhisel, R. I., and Thom, W. O., Laboratory spoil tests and interpretation for assessing fertility levels, *Reclamation News and Views,* Coop. Ext. Serv., University of Kentucky, Lexington, 3, 1, 1985.
63. Spry, M. J., Revegetation Research on Hard Rock Mining Disturbances in North-Central Montana, M.S. thesis, Montana State University, Bozeman, 1986, 68.
64. Williamson, N. A., Johnson, M. S., and Bradshaw, A. D., *Mine Waste Reclamation,* Mining Journal Books Ltd., London, 1982, 34.
65. Kay, B. L., Mulch and chemical stabilizers for land reclamation in dry regions, in *Reclamation of Drastically Disturbed Lands,* Schaller, F. W. and Sutton, P., Eds., American Society of Agronomy, Madison, WI, 1978, 467.
66. Theisen, M. S., Cost-effective techniques for successful erosion control, in Proc. High Elevation Reveg. Workshop No. 8, Kemmerer, W. R. and Brown, L. F., Eds., Col. Water Resour. Res. Inst. Inf. Ser. No. 59, Colorado State University, Fort Collins, 1988, 240.

67. Proteau, J. T., Revegetation principles, practices, and problems associated with the development of a mountain resort, in Proc. High Elevation Reveg. Workshop No. 8, Kemmerer, W. R. and Brown, L. F., Eds., Col. Water Resour. Res. Inst. Inf. Ser. No. 59, Colorado State University, Fort Collins, 1988, 291.
68. Kill, D. L. and Foote, L. E., Comparison of long- and short-fibered mulches, *Trans. Am. Soc. Agric. Eng.*, 14, 942, 1971.
69. Kay, B. L., Mulch and chemical stabilizers for land reclamation in dry regions, in *Reclamation of Drastically Disturbed Lands*, Schaller, F. W. and Sutton, P., Eds., American Society of Agronomy, Madison, WI, 1978, 467.
70. Golueke, C. G., *Biological Reclamation of Solid Waste*, Rodale Press, Emmans, PA, 1977, 27.
71. Reclamation Research Unit, *Rapid Soil Stabilization, A Roadside Development Summary Report*, prep. for U.S. Department of Transport and MT Department of Highways, Montana State University, Bozeman, 1981, 11.
72. Reclamation Research Unit, *Rapid Soil Stabilization, A Roadside Development Summary Report*, prep. for U.S. Department of Transport and MT Department of Highways, Montana State University, Bozeman, 1981, 11.
73. Schuman, G. E., Taylor, E. M., Meining, J. L., and Belden, S. E., Role of organic and inorganic amendments in reclaiming bentonite spoils, in Proc. Symp. Evolution Abandoned Mine Land Tech., Riverton, WY, 1989, 269.
74. Mannering, J. V. and Meyer, L. D., The effects of various rates of surface mulch on infiltration and erosion, *Soil Sci. Soc. Am. Proc.*, 27, 84, 1963.
75. Meyer, L. D., Wischmeier, W. H., and Foster, B. R., Mulch rates for erosion control on steep slopes, *Soil Sci. Soc. Am. Proc.*, 34, 928, 1970.
76. Thornburg, A. A., Plant Materials for Use on Surface-Mined Lands in Arid and Semiarid Regions, S.C.S. TP-157, U.S. Department of Agriculture, Washington, D.C., 1982, 8.
77. Reclamation Research Unit, *Rapid Soil Stabilization, A Roadside Development Summary Report*, prepared for U.S. Department of Transport and MT Department of Highways, Montana State University, Bozeman, 1981, 13.
78. Darling, A. R., The Effects of Native Hay Mulch on Stabilization and Revegetation of Strip Mined Lands in Southeastern Montana, M.S. thesis, Montana State University, Bozeman, 1983, 74.
79. Schuman, G. E., Taylor, E. M., Rauzi, F., and Howard, G. S., Standing stubble versus crimped straw mulch for establishing grass on mined lands, *J. Soil Water Conserv.*, 35, 25, 1980.
80. Martin, J. S., The erosion control and revegetation mat . . . product and application, in Proc. High Elevation Reveg. Workshop No. 7, Schuster, M. A. and Zuck, R. H., Eds., Col. Water Resour. Res. Inst., Inf. Ser. No. 58, Colorado State University, Fort Collins, 1986, 120.

81. Kay, B. L., Cost and effectiveness of mulching practices, in Proc. High Elevation Reveg. Workshop No. 6, Colbert, T. A. and Cuany, R. L., Eds., Col. Water Resour. Res. Inst. Inf. Ser. No. 53, Colorado State University, Fort Collins, 1984, 170.
82. Meek, B., Graham, L., and Donovan, T., Long-term effects of manure on soil nitrogen, phosphorus, potassium, sodium, organic matter, and water infiltration rate, *Soil Sci. Soc. Am. J.*, 46, 1014, 1982.
83. Diaz, L. F., Savage, G. M., and Golueke, C. G., *Resource Recovery from Municipal Solid Wastes, Vol. II, Final Processing,* CRC Press, Boca Raton, FL, 1982, 95.
84. Brandt, C. A. and Hendrickson, P. L., *Use of Composts in Revegetating Arid Lands,* PNL—7833, Battelle Pacific Northwest Laboratory, Richland, WA, 1991, 31.
85. Brandt, C. A. and Hendrickson, P. L., *Use of Composts in Revegetating Arid Lands,* PNL-7833, Battelle Pacific Northwest Laboratory, Richland, WA, 1991, 11.
86. Hill, R. D. and Montague, A., The Potential for Using Sewage Sludges and Compost in Mine Reclamation, paper presented at 3rd Natl. Conf. Sludge Manage., Disposal, Util., Miami, FL, 1976, 5.
87. Oyler, J. A., Reclamation of site near a smelter using sludge/fly ash amendments: herbaceous species, in Mine Drainage and Surface Mine Reclam. Proc. of a Symp. Vol. II, Bur. of Mines Inf. Circ. 9184, U.S. Department of the Interior, Pittsburgh, PA, 1988, 22.
88. Sopper, W. E., Reforestation of a zinc smelter superfund site, in Mine Drainage and Surface Mine Reclam. Proc. of a Symp. Vol. II, Bur. of Mines Inf. Circ. 9184, U.S. Department of the Interior, Pittsburgh, PA, 1988, 68.
89. Halderson, J. L. and Zenz, D. R., Use of municipal sewage sludge in reclamation of soils, in *Reclamation of Drastically Disturbed Lands,* Schaller, F. W. and Sutton, P., Eds., American Society of Agronomy, Madison, WI, 1978, 355.
90. Williams, D. E., Vlamis, J., Pukite, A. H., and Corey, J. E., Trace element accumulation, movement, and distribution in the soil profile from massive applications of sewage sludge, *Soil Sci.*, 129, 119, 1980.
91. Stover, R. C., Sommers, L. E., and Silviera, D. J., Evaluation of metals in wastewater sludge, *J. Water Pollut. Control Fed.*, 48, 2165, 1976.
92. U.S. Environmental Protection Agency, Municipal Sludge Management: Environmental Factors, MCD-28, Tech. Bull., Off. of Water Prog. Operations, U.S. Environmental Protection Agency, Denver, CO, 1977, 19.
93. Dowdy, R. H., Goodrich, R. D., Larson, W. E., Bray, B. W., and Pamp, D. E., Effects of Sewage Sludge on Corn Silage and Animal Products, Off. of Res. and Dev. EPA- 600/2–84–075, U.S. Environmental Protection Agency, Cincinnati, OH, 1984, 3.
94. Anderson, T. J., Barrett, G. W., Clark, C. S., Elia, V. J., and Majeti, V. A., Metal concentrations in tissues of meadow voles from sewage sludge-treated fields, *J. Environ. Qual.*, 11, 272, 1982.

95. Dressler, R. L., Storm, G. L., Tzilkowski, W. M., and Sopper, W. E., Heavy metals in cottontail rabbits on mined lands treated with sewage sludge, *J. Environ. Qual.*, 15, 278, 1986.
96. Williamson, N. A., Johnson, M. S., and Bradshaw, A. D., *Mine Waste Reclamation*, Mining Journal Books Ltd., London, 1982, 76.
97. Farmer, E. E., Richardson, B. Z., and Brown, R. W., Revegetation of Acid Mining Wastes in Central Idaho, Res. Pap. INT-178, Interm. For. and Range Exp. Sta., U.S. Department of Agriculture, Ogden, UT, 1976, 16.
98. Maclean, A. J. and Dekker, A. J., Lime requirement and availability of nutrients and toxic metals to plants grown in acid mine tailings, *Can. J. Soil Sci.*, 56, 27, 1976.
99. Schuman, G. E. and Meining, J. L., Short Term Gypsum Effects on Sodic Bentonite Spoils, in press, 1993.
100. Dollhopf, D. J., Rennick, R. B., and Smith, S. C., *Long-term Effects of Physicochemical Amendments on Plant Performance at a Bentonite Minesite in the Northern Great Plains*, Reclam. Res. Unit Publ. No. 8802, Montana State University, Bozeman, 1988, 67.
101. Young, S. A. and Rennick, R. B., *Establishment of Seeded Species on Coal Surface Minesoils Following Temporary Irrigation*, Agric. Exp. Sta. Res. Rep. 200, Montana State University, Bozeman, 1982, 16.
102. Young, S. A. and Rennick, R. B., *Establishment of Seeded Species on Coal Surface Minesoils Following Temporary Irrigation*, Agric. Exp. Sta. Res. Rep. 200, Montana State University, Bozeman, 1982, 56.
103. Powell, K. B., Vincent, R. B., DePuit, E. J., and Parady, F. E., Temporary irrigation for revegetation of mined lands under cool desert conditions, in *Proc. 4th Biennial Symp. Surf. Min. Reclam. Great Plains*, Reclam. Res. Unit. Publ. No. 8704, Montana State University, Bozeman, 1987, D-4–1.

4

Seeding and Planting

The final and most visible phase of the restoration program is revegetation. In a narrow sense, this term may be restricted to seeding and/or planting various species of plants on degraded soil, but as this text emphasizes, such activities cannot be separated entirely from the other aspects of reclamation. The objectives guiding plant community development on all disturbances revegetated with any plant species under almost any law or regulation are identical: namely, erosion control and forage for animal consumption.

Before the vegetation specialist can place seed in the ground, the soil resource on which all plants depend must be reconstructed. Once the soil resource has been replaced, the reclamationist can create a new plant community on this material. However, what plant community will best fill the void created by destruction of the vegetation cover on the site? Pioneer species usually begin the process of rehabilitation. These opportunistic species are replaced by a series of plants in a process called plant succession. Succession leads to the development of a plant community somewhat like the predisturbance plant community, but this may require many years. Although pioneer plants could be permitted to initiate reclamation on the disturbance, they are characteristically weedy, less palatable to livestock and wildlife, and exert less control over erosion than later seral species. In addition, man has accidentally and intentionally introduced weedy species into every community on the globe. Many of these plants act as pioneers but interfere with or arrest succession before it reaches the stage most reclamationists hope to attain. The revegetation scientist should strive to develop a plant community that is more mature than the classic pioneer stage to ensure that undesirable weedy plant species do not become established on the degraded site.

The first manipulations to a site so that it will be capable of supporting later seral plant species are recontouring and topsoiling. These two processes alone move the successional process from the primary level to secondary succession. The soil maturation that takes place under pioneer species is not necessary when surface soil is available. Topsoil is generally mature enough to support mid-seral stages of plant communities at degraded sites. Successional change can be further accelerated by seeding selected later-successional species. This action completely avoids the first stages of plant succession. Plant species available at commercial seed houses are usually mid-seral species and complement the successional progress introduced by topsoil application.

The rehabilitation scientist may also seed late successional or climax species, but habitat requirements for such communities, those beyond the early to mid-seral stages, cannot be met in most disturbed site soils. During topsoil stripping and respreading, properties of the soil are damaged. Complex nutrient cycles, mycorrhizal associations, and even simple phenomena such as surface litter distribution, absorption of solar radiation, and surface soil microtopography are modified after disturbance. These changes may prohibit the successful development of late seral species on disturbed sites. If seeds for late successional stages are placed in disturbed soils, the revegetation program may be a failure simply because the root media and its accompanying flora and fauna, or other site characteristics, were not adequate to support the seeded species.

The Surface Mining Control and Reclamation Act (PL 95–87, August 3 1977) specifically uses the terms *previous use, native,* and *diverse* in its discussion of revegetation goals. This has been interpreted to mean that coal mining disturbances are to be returned to native grasslands, shrub lands, or forests unless the land was in agricultural production before mining. Recent revegetation programs on disturbances other than coal mining have begun to emphasize the use of native vegetation. There are major exceptions, but these are on unique reclamation problems such as tailings or acid materials.[1] Even on these sites, the concepts of previous use and diversity influence revegetation.

Vegetation to be seeded or planted on degraded lands may be native or introduced. Native plants are those endemic to an area. Introduced plants, which are those that have been brought into an area by man, may be naturalized or exotic. Naturalized plants are those adapted to the area of introduction. For example, crested wheatgrass *(Agropyron cristatum)* is well adapted to the Great Plains but was brought to this area from Russia. Kentucky bluegrass *(Poa pratensis)* also performs very well in riparian habitats of the northern Plains, but it was probably brought to this region by early settlers. Both plants are considered naturalized because they survive and reproduce well in this geographic region. When the performance of other introduced plants on the plains is not known, the plants are referred to as exotic. While the short-term performance of an exotic plant may be encouraging, long-term success is much more important for land revegetation. For example, the short-term performance of Eucalyptus trees (*Eucalyptus camaldulensis* and *E. microtheca*) planted on tailings in Arizona is very good, but the reproduction and long-term performance of these species is not certain in this environment and, therefore, the plants must be considered exotic. This does not mean the vegetation is undesirable or bad nor that the plants are especially good, it simply means that the long-term desirability or persistence of the species is unknown.

The term *diversity* is listed by SMCRA as one of the attributes of plant communities on revegetated coal minesoils. Although this law was passed in 1977, the question of what was meant by this term has not been defined. Whittaker speaks of two types of diversity, alpha and beta.[2] Alpha diversity is defined as species richness or the diversity within a community. That is, the number, frequency, and distribution of species within a plant grouping in a specific habitat. Beta diversity is a measure of diversity between habitats. If plant communities are mapped as small, discrete units, alpha diversity is low but beta diversity is high. If plant communities are conceived as large expanses of similar vegetation, alpha diversity is high but beta diversity is low. Most reclamation specialists are inclined to view the requirement of SMCRA as alpha diversity; however, the successional changes in plant groups and the evolution of different communities across the landscape (beta diversity) is receiving growing consideration. Numerous species are therefore included in coal minesoil plant seedings to increase the number of species in the community and to permit the establishment of different communities on different micro- or macrohabitats without a separate seeding for each community.

SEEDBED PREPARATION

After recontouring, but prior to actual placement of seed or plants into the soil, certain activities must take place on the disturbed site to facilitate seeding and improve the probability of seeding success. In almost all examples of large revegetation efforts some type of surface soil is placed over the subsoil. It is a rare revegetation attempt today that is seeded directly in spoil material. The topsoil (or coversoil, since the two terms are used interchangeably in this text) may be from storage piles constructed days, months, or years previously, or the material may have been picked up, hauled to the area to be revegetated, and placed on the subsoil in one operation. In either example, the surface covering needs some type of manipulation to level the seedbed, break up large clods of soil, and reduce the soil seedbank of competitive plants. If the topsoil is not immediately seeded, any soil crusts that form must be broken up. These types of activity are called tillage.[3] Tillage is the operation of plows, chisels, harrows, disks or other implements on the seedbed to prepare a soil that is loose enough to permit the growth of the seed root and infiltration of water yet firm enough to ensure close soil-seed contact.

Before tilling, soil samples are usually collected from the seedbed for nutrient analyses and determination of amendment needs. On most semiarid disturbances such tests are unnecessary if analyses indicated that the surface soil was of good quality, if the topsoil was stripped and placed directly on recontoured, nontoxic materials or stored for only a short period of time (less than one growing season), and native species were seeded. Most native grassland species of North America evolved in a nutrient-poor environment. Under the conditions stated above, seeds of these endemic species will germinate, establish, grow, and reproduce without the addition of fertilizers or amendments to the seedbed. Addition of unnecessary amendments to such sites only contributes to increasing costs of reclamation without a concurrent increase in the quality of the new vegetation.

If the site is to be drill seeded, the bed should be smooth to permit the proper placement of seed in the soil. This practice encourages maximum germination and establishment of large-seeded species. A rough surface is necessary for broadcast seeding to ensure proper seed placement. Broadcast seeding encourages the germination and establishment of small-seeded species.

SEED MIXES

Every disturbance is unique and seed mixes should be tailored to the soils, climate, environmental setting, proposed land use, and plant community desired on the site. The revegetation specialist must design each seed mix specifically for the site to be seeded. Thus, it is common knowledge that cool-season grasses are very competitive in the northern Great Plains, but what particular cool-season grasses perform best on highway shoulders in eastern Montana? Warm-season grasses are very successful in the southern Plains, but what species of warm-season grasses should be seeded on waste rock dumps around Tucson, Arizona? Furthermore, should warm-season grasses be seeded on the northern Great Plains and cool-season grasses on the southern Plains? This questioning can be extended to other life forms. Should forbs be included in the seed mix, should trees and shrubs be seeded?

Reclamation scientists attempt to reconstruct plant communities on degraded soils that match or exceed the potential productive capacity of the site prior to disturbance. It has long been assumed that the predisturbance plant community represented the maximum long-term potential of the site. Ecological research has revealed that many plant communities in semiarid to arid areas have been so degraded by past or present abuses that they are not reaching their biological potential.[4] On such sites the revegetation specialist must utilize adjacent plant communities or exercise educated judgment to determine appropriate species for seeding.

Seeds of the plant species that dominate the plant communities identified on the site prior to disturbance or that the revegetation specialist has determined are appropriate for seeding on this disturbance may not be available in quantities sufficient to meet the needs of the program. Substitutions, therefore, are forced upon revegetation planners. To maintain the integrity of the new community, postdisturbance grass, forb, and shrub species in the seeding should be matched with similar life forms that were present in the predisturbance plant community. This requires the planning of complex seed mixtures, seeding techniques, and vegetation management programs.

Seed mixes composed of numerous species and life forms are the rule on large disturbances today. Unfortunately, a few aggressive species may dominate the landscape after one or two growing seasons. This dominance is exacerbated by uniform slope, exposure, and topsoil quality and depth on most large disturbances. The diversity inherent in these parameters on the natural landscape is not found on rehabilitated lands. Without the diversity of macro- and microhabitats found on native grasslands a plant community cannot be developed on degraded soils as diverse as that found on the site prior to disturbance. Nevertheless, the revegetation specialists must attempt to produce as diverse a community as possible by manipulating species selection, seed rates, seeding methods, and other techniques available to the program.

Plant Species Selection

Within the limitations imposed by past abuse, if a disturbed site is to be returned to its previous use the seed mix should reflect predisturbance vegetation and land use. To meet this requirement, the selection of plant species to be seeded on a site must be based upon at least five factors:

1. Habitat
2. Life forms expressed in the predisturbance plant community
3. Seed availability
4. Species usefulness for the designated postdisturbance land use
5. Regulatory restrictions

The growth requirements of each member of a seeded plant community must match the texture, pH, salinity, nutrient levels, and all of the chemical and physical characterists of the rootzone materials. Seeded plants must also meet aspect, elevation, precipitation, and temperature limitations of the area. For example, the northern Great Plains is dominated by spring precipitation patterns. As a general rule, grasses and forbs seeded in this region must possess the genetic complement of the cool-season species that permits them to utilize early growing-season rainfall. While warm-season grasses are widely distributed over the northern Great Plains they are found in low numbers in most habitats. They occur in greater numbers on coarse-textured hillsides or where south or southwest exposures create habitats in which cool-season grasses perform very poorly. Grasses and forbs seeded in the southern Plains cannot depend on spring precipitation but usually receive some summer moisture. They must quickly absorb the moisture from and respond to these summer precipitation events. Therefore, warm-season species are common in the southern Plains. These two large groups of grasses form the basis for revegetation of disturbances throughout North America, but representatives of other lifeforms serve important functions in the reclaimed habitat and contribute to diversity on the revegetated site.

In selecting a group of species for seeding, the vegetation specialist should know not only how the habitat parameters affect each seeded species or cultivar, but also how subtle variations of combinations of these parameters influence vegetation. An example illustrates this point. Soil texture and precipitation both influence the dryness of a site. Interactions between soil characteristics (texture) and climatic factors (precipitation) with slope gradient and aspect create additional moisture complications for the revegetation scientist. Thus, a south-facing 3:1 slope in a 30-cm (12-in.) precipitation zone is effectively much drier than the precipitation data would indicate. A silty-clay soil with a high sodium content is more difficult to revegetate than a high-sodium sandy-loam soil. The revegetation scientist must recognize these combinations of characteristics in selection of plant species for seeding on any disturbance. Species selection should not be based on mean soil or climatic factors alone but on as many sets of data as can be economically gathered from the specific disturbed site.

It is common to group plant species into life forms such as perennial grasses, annual grasses, perennial forbs, annual or biennial forbs, and shrubs or trees. These categories may be further broken into additional subgroups. For example, perennial grasses and most forbs may be divided into cool- and warm-season species. Trees and shrubs may be deciduous or evergreen. Terminology such as cool season perennial grasses, annual forbs, and deciduous trees are commonly used in descriptions of disturbed land plant communities.

Most reclamation legislation like SMCRA requires that the plant community on the reclaimed soil should mimic the life forms, if not the species, found in the predisturbance inventory. Unfortunately, seed will not be available for all the species identified in the inventory. Prior to the 1970s very few native seed sources were available to specialists planning revegetation. Since that time the Plant Materials Centers and commercial seed houses have developed numerous varieties or cultivars of many native grass, forb, and shrub species for seeding or planting on disturbed lands. Many of these varieties represent species or life forms found in the predisturbance site inventory. However, when commercial cultivars do not represent exact species found in the predisturbance inventory, they often may be substituted into the seedmix for plant species or life forms identified in the predisturbance vegetation community but for which seed is not commercially available. Rosana western wheatgrass (*Agropyron smithii* var. Rosana), Critana thickspike wheatgrass (*Agropyron dasystachyum* var. Critana), and Lodorm green needlegrass (*Stipa viridula* var. Lodorm) are three common examples of cool-season grasses available for seeding on the northern Great Plains.

Seed should be available for some species within each life form, but this does not mean that appropriate species or varieties can be seeded. Several cultivars or ecotypes are available for many of the plant species used in land rehabilitation programs, but seed may not be available for ecotypes adapted to the habitat of the disturbance. For example, several cultivars of western wheatgrass are grown commercially and their seed is sold for use in parts of the northern Great Plains (Barton, Rodan, Rosana, etc.), but no cultivars of tufted hairgrass *(Deschampsia caespitosa)* are available for seeding in the same region. When selecting the species, cultivar, or variety for a particular disturbance, the reclamation scientist must choose the species and/or ecotype that is most appropriate for the environmental, physical, and chemical characteristics of the site.

When commercial cultivars of a plant are not available, seed or seedlings of wild plants are often collected and planted. For frequently seeded species, a seed crop is often produced from wild seed planted in pure stands. In all instances, when wild seed is collected for seeding or when wild seed is planted to grow a seed crop, the original seed or seedlings are selected from a plant population that evolved on a particular soil under a unique environmental regime. When such an ecotype is utilized in the rehabilitation of a disturbance, the original seed source should be close to the site of revegetation use. It has been suggested that an ecotype can perform satisfactorily only if it is seeded at approximately the same elevation, exposure, aspect, and if it is moved no more than 400 to 480 km (250 to 300 mi) north or 160 to 240 km (100 to 150 mi) south of the point of origin of the seed.[5] East or west movement may be greater or less, depending on changes in elevation and precipitation. While there has been little research to confirm this generalization, much experience and many observations by revegetation specialists support it. Even this limited movement depends on equivalence of the soils, temperature extremes, and moisture at the original source of the seed and the site to be seeded. These limitations do not apply to commercial cultivars because these plants were genetically selected for specific attributes within a specified range of soil and climatic conditions. A seed crop for a cultivar produced by one of the plant materials centers may be grown in any environment, will retain its genetic complement, and will perform adequately under the limitations originally described for the cultivar.

Seeded species should contribute to the attainment of the land use goal for the site. At almost all sites these include soil stabilization, erosion control, and soil development. In general, on semiarid and arid rangelands the land use goal is livestock grazing and wildlife habitat. Under the land use goal — livestock grazing — plant species selected for seeding should meet the needs of the species of animals that will be feeding on and inhabiting the site. Season of use, palatability, and nutritional value of the forages must also be factors in plant selection. If wildlife habitat was included as a specific goal of the proposed land use, food and cover for the wildlife species anticipated on the site must be addressed in the seed mix. Species selected for seeding on lands destined for other post-mine uses must meet other criteria, but they must still address the soil stabilizing, erosion control, and soil development goals of the site.

Multiple land use goals dominate reclamation programs. Diverse plant communities meet these goals better than simple mixtures of a few grass and forb species. Surface coal mines, and to a lesser extent all mining and large-scale surface disturbances, have multiple land use goals. Therefore, rehabilitation experts are confronted with the need, if not a legal requirement, to produce a diverse plant community on the post disturbance landscape.

Diverse plant communities offer mixed diets and habitat for livestock and wildlife. Forage production tends to be maintained throughout the growing season because of variable life forms, species, and populations in the community. The variety of rooting patterns expressed by multiple species and life forms encourages greater total production as these forms utilize a larger component of the soil water and nutrient supply. More niches are filled by the greater number of adapted species. Thus, diversity is clearly desirable for new plant communities.

Regulatory restrictions affect revegetation efforts on many disturbances. Most notable are the native plant cover and production requirements on coal surface mines in the United States. Many species or varieties of introduced grasses are commercially available. Some of these are adapted to the soils and environment of the semiarid or arid West. These naturalized species possess characteristics that permit them to persist on disturbed soils in this region. For the vast majority of naturalized species there are other characteristics that reduce their desirability as revegetation species. They may be overly aggressive, nonpalatable during parts of the year, or require fertilization. Regulatory restrictions control the seeding of these species on coal minesoils, but not on other types of disturbances.

Seed Rates

The number of seeds placed in a unit area of soil is called the seed rate. The total seed rate is the sum of the individual species' seed rates. Seed rates are normally expressed as the number or weight of seeds per hectare (acre). Numerous different seed rates for the same species are suggested in the literature. Part of the confusion is caused by authors who specify seed rates for pure stands and others who suggest rates for mixed seedings. A second source for this confusion is the varying contents of viable seed and nonseed material in the samples used in seeding research. This problem may be corrected if the percent pure live seed (PLS) in a container of seed is specified. Pure live seed is calculated by the formula:

$$\%\text{PLS} = \frac{\%\text{Germination} \times \%\text{Purity}}{100} \qquad [4]$$

where Germination = % of viable seeds in a unit weight

Purity = 100% − % trash − % weed seed

All seed should be purchased on the basis of PLS and require an accompanying certification.

Most states have a seed certifying agency and seed growers may adopt certification programs. Certification of a container of seed by a grower assures the customer that the seed is correctly identified and genetically pure. The state agency responsible for seed certification sets minimum standards for mechanical purity and germination for each species of seed. When certified, a container of seed must be labeled as to origin, germination percentage and date of the test, the percentage of pure seed (by weight), other crop and weed seeds, and inert material. The certification is the consumer's only guarantee that he is purchasing good quality seed. Because of the definition of certification, only cultivars released by the Plant Materials Centers or developed by seed companies may be certified. Seed collected from native plants may not be certified.

The seed rate may be calculated from an expected field emergence for each species and the desired number of plants per unit area. For purposes of calculation, field emergence for small seeded grasses and forbs is assumed to be around 52% if germination is greater than 80%, and 33% if germination is between 60 and 80%.[6] The Soil Conservation Service recommended 20 PLS per 0.09 m^2 (20 PLS per ft^2) as a minimum number of seeds on drill seeded soils.[7] Twenty pure live seeds per 0.09 m^2, with an expected field emergence of 50%, produces approximately ten plants on the seeded area. On favorable seed beds (i.e., good loam soils, north or east facing, gentle slopes [5:1 or less], and good moisture regimes) the seed rate may be lowered. Harsh sites receive higher seed rates and broadcast seeding usually requires twice the number of seeds recommended for drill seeding.

One factor influencing seed rates that is often overlooked is competition between seeded and weedy species. The number of plants actually germinating and establishing on a site must be sufficient to inhibit the development of weeds on the seed bed. While the number of seeds drilled on good sites may be reduced below 20 seeds per 0.09 m^2, it must not be reduced below the number needed to exclude weeds from the site. This fine tuning of seed rates must be site-specific and can be determined only by experimentation in each different habitat at a disturbance. In the absence of experimental evidence to suggest otherwise, the recommendation of 20 PLS per 0.09 m^2 should be used as a general guide for drill seeding on seedbeds composed of good quality topsoil.

In addition to expected field emergence, the competitive ability of each species must be evaluated to determine species seeding rates. In general, grasses are more competitive than forbs, forbs are more competitive than shrubs, and shrubs are more competitive than trees. Among grasses, the naturalized species are usually more aggressive than native species. Within native species on the northern Great Plains, cool-season grasses are more competitive than warm-season grasses.

SEEDING TECHNIQUES

The primary concern of seeding is to place the seed in the soil at the depth most favorable for its germination and successful establishment. The optimum depth of seed placement differs for each plant species. In general, the smaller the seed, the more shallow the placement in the soil. Furthermore, light stimulates germination in some species, darkness stimulates it in others;[8] therefore, some seeds should be buried while others should be placed on the soil surface. When a seed germinates, the embryonic root emerges from the seed coat. This root must absorb water for the young plant. If the air around the root is moist and the root makes contact with damp soil, successful establishment is likely, but if the root is some distance from the soil, as little as 2.5 mm (0.1 in.) in some small-seeded species, the seedling may die before the root can absorb adequate water for the rapidly developing plant. Because of these differing germination requirements, the best conditions for all plants cannot be met by any single seeding technique. However, economics dictate that only one or two seeding passes be made on a disturbed soil. A compromise, therefore, must be made and an average set of conditions met by the final seeding process.

Drill Seeding

This type of seeding is used for large-seeded species. Examples include plants such as the wheatgrasses, most legumes, and large-seeded shrubs like fourwing saltbush. Drill seeding requires an implement that places the seed at a specified depth in the soil (Figure 28). Typically, on a drill seeder two disks open a furrow in the prepared seedbed to a predetermined depth, seeds are metered or dropped through a tube into the furrow, and a set of disks or wheels push dirt over the seeds. The final step in this process incorporates a cultipacker

Figure 28 Drill seeder.

(Figure 29), which is a soil compacting device that rolls over the soil and firms the seed bed. Drill seeding guarantees seed placement at a predetermined soil depth and good soil-seed contact. It is the preferred seed dispersal mechanism when seedbeds are smooth and large seeds are used.

The optimum depth for placement of seed varies for each species. In general, larger seeds may be seeded deeper in the profile. Seeds such as western wheatgrass are usually seeded 2.5 to 4 cm (1 to 1.5 in.) deep, whereas a small seed such as that of timothy *(Phleum pratense)* should be placed only 1.0 to 1.5 cm (0.5 in. or less) into the soil.

One of the desirable aspects of drill seeding is the potential for optimization of seed placement to maximize seed contact with water. Seeding depth, therefore, varies with soil water-holding capacity, soil texture, site exposure, aspect, and other characteristics that influence soil moisture. Drills should be set at deeper depths in light sandy soils or on sunny south or southwest aspects. On north- or northeast-facing slopes, in moist areas, or in finer-textured soils with elevated water-holding capacity, seeds may be planted at shallower depths. Seed companies, the Soil Conservation Service, and local county agents will make seeding depth recommendations in different areas.

Unfortunately, the young plants from drill seeders germinate in rows that resemble a crop rather than a native plant community (Figure 30). A more important drawback of drill seeding is the separation

Figure 29 Cultipacker behind a disk.

Figure 30 Drill-seeded plant community on a minesoil.

of seeds of various sizes in the seed container prior to deposit in the soil. Very small seeds vibrate to the bottom of the seed box and fall from the box faster than larger seeds. The segregation of seeds by size or weight can be overcome by addition of a carrier to the seed mix. A carrier is any substance that dampens vibrations in the seed boxes and prevents seeds from separating according to size. Almost any dry organic material may be used as a carrier (e.g., ground coconut hulls, ground corn cobs, or something as innocuous as sand).

Drill seeders usually have multiple seed boxes — often, one box for each row. The placement of seeds into separate boxes on the basis of their size is practiced. Because each seed box drops seed into a unique row, the placement of different seeds into separate seed boxes creates a plant community consisting of rows of different plant species.

Fuzzy or hairy seeds, or seeds with long awns (the so-called trashy seeds) also create problems in drill seeders. These seeds form large bunches that interfere with the fall of individual seeds from the boxes into the seeding tubes. Again, some type of carrier may be added to the box to control the flow of seeds from the hopper. Drill seeding is more energy intensive and permits seeding at a limited number of different depths, but it ensures close soil-seed contact and a firm bed over the seed, and maximizes the probability of successful germination and plant establishment.

Broadcast Seeding

Any method of seed dispersal that drops the seed on the ground instead of placing it in the ground is called broadcast seeding. Broadcast seeding implements range from mechanical devices that indiscriminately throw seed from a container onto the soil surface (Figure 31) to complex seeders that combine some attributes of drill seeders with the dropping of seeds on the soil surface (Figure 32). Because the seed is deposited on the soil surface and not placed in the soil, some sort of device (e.g., a chain) is usually dragged over the site after seeding to knock some of the seed into cracks and crevices in the soil. This may be followed by rollers or cultipackers to press the seed into the soil for good soil-seed contact. This technique provides variable-depth seeding so necessary with multiple seed sizes and is the preferred seeding method for small-seeded species.

Broadcast seeding provides a randomly dispersed plant community that is more like a native plant community than drill-seeded planting (Figure 33). However, since many seeds germinate on or near the soil surface, they often desiccate before the root can absorb sufficient water to maintain the seedling. Other similarly placed seeds never germinate or establish because they are consumed by seed-eating rodents or birds. The number of seeds, therefore, must be greater than that used in drill seeding. A doubling of drill seed rates for broadcast seeding is normally recommended.

Hydroseeding is a form of broadcast seeding in which the seeds are dispersed in a liquid under pressure. The liquid is usually water,

Figure 31 Broadcast seeder or fertilizer spreader.

Figure 32 Brillion broadcast seeder.

Figure 33 Broadcast-seeded plant community on a minesoil.

although it may include fertilizer or other additives. Seeds should not be included with hydromulch and/or a tackifier because many of the seeds are suspended above the soil as the hydromulch and tackifier dry. These may germinate, but they dry out and die before they can establish. Hydromulch may be applied over hydroseeded sites with or without a tackifier. Hydroseeding is very effective on steep slopes or rough-textured materials than cannot be drill seeded (e.g., waste rock and roadcuts).

Aerial seeding, also a form of broadcast seeding, is the dispersal of seed by fixed-wing aircraft or helicopters. If properly carried out, good ground cover can be attained with this form of seed distribution (Figure 34). Aerial seeding is cost effective when large areas are to be seeded. It overcomes the problem of site accessibility and the need for access roads. If seed rates and seasonal conditions are carefully evaluated, large amounts of seed (and amendments) may be distributed in a timely manner without damage to existing vegetation on the site.[9] Aerial seeding is used on very rough, inaccessible terrain. It is commonly used in mountainous regions after fires.

A unique combination of drill and broadcast seeding is being carried out at certain mines in Wyoming. On a modified drill seeder alternate seed boxes are loaded with seed to be drilled and seed to be broadcast. The seed can be drilled and broadcast in a single operation when two devices are removed from the hoppers containing broadcast seed: (1) the disks that open the furrows, and (2) the wheels that cover the seed openings or rows. The cultipackers not only firm the drilled seed bed but also press the broadcasted seed into the seed bed.

SEEDING LEGUMES

Legumes are included in almost every seed mix. They are the most common group of nitrogen-fixing species. Other higher plants and

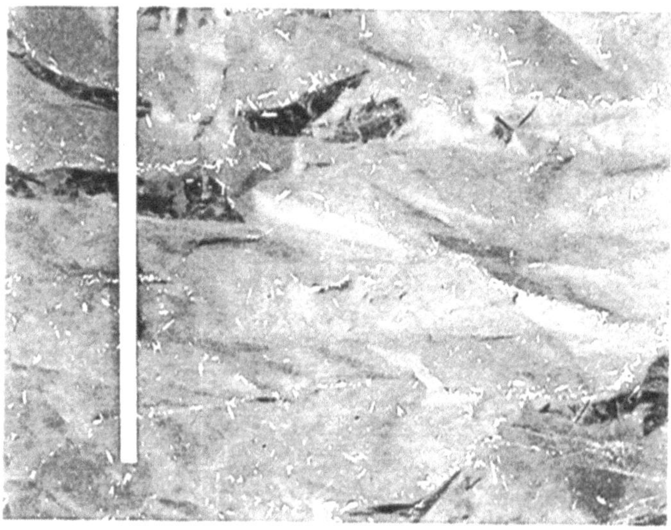

Figure 34 Seed dispersal by fixed wing aircraft.

some bluegreen algae also fix nitrogen, but the contribution of these species is small compared to that of the legumes. All of these nitrogen-fixing organisms make an invaluable contribution to the reestablishment of the nitrogen cycle in disturbed soils. The nitrogen component added to the ecosystem by legumes is a result of their ability to form symbiotic relationships with certain strains of bacteria that convert atmospheric nitrogen into plant-available nitrogen. These nitrogen-fixing bacteria live and reproduce in nodules on the roots of nitrogen-fixing plants. Bacteria and higher-plant relationships are species specific. The presence of the appropriate bacteria in the soil should not be left to chance. Legume seed should be inoculated with the species of bacteria appropriate for the higher plant species prior to seeding to ensure maximum nodulation and nitrogen fixation.

Native legumes fix smaller quantities of atmospheric nitrogen than most agronomic species, therefore almost all seed mixes include one or more agronomic legumes. These plants make a significant contribution to soil nitrogen levels, but they should not be considered replacements for native legumes. Agronomic species are gradually displaced from the community by more aggressive plants. Native legumes, however, reproduce and perpetuate the species on the disturbance, since they are well adapted to the climate and soils of the area.

SEEDING TEMPORARY STABILIZING SPECIES

Pioneer species germinate on degraded sites within one or two growing seasons after disturbance with no input from man. These opportunists grow very rapidly and need no protection from wind, sun, or high temperatures. On the other hand, perennial grasses, forbs, and shrubs are slow growing. Young plants of these life forms need protection from sun, wind, and high temperatures, at least during their first growing season. Pioneers provide to highly unstable and erodible soils the environmental protection and immediate stability that is necessary for the germination and establishment of slower growing, later seral species. Unfortunately, pioneer species are usually weedy, nonpalatable, and some persist for several years. They are not desirable long-term plant species for degraded sites. So-called nurse crops or companion crops have been used on disturbed soils to provide the immediate soil stabilization and protection for seeded species that would be provided by pioneer species in the natural habitat.

Russian thistle *(Salsola kali)* is an example of a pioneer species that germinates on disturbed soils of all types throughout western North America. It functions as a nurse crop on many degraded sites in this region. It shades perennial grasses and forbs growing under its canopy and protects them from wind and high temperatures. This species commonly germinates on minesoils, pipeline corridors, and highway rights-of-way in this region.

Some mines mow the temporary stabilizing species (or the Russian thistle) at the end of the first growing season, others permit the litter to remain standing. In either case, these plants do not germinate or they appear in very reduced numbers during the second growing season. Perennial plants that have established strong root systems under these annuals develop rapidly in their second year. This rapid growth and development in year two is thought to be a result of both well-developed root systems and release from the competition of the annual nurse crop.

When temporary stabilizing species or companion crops are seeded with perennial vegetation on disturbed soils they are referred to as stubble mulch crops. They help control water and wind erosion plus annual grass development. The nurse crops lower soil surface temperatures and reduce water loss from the soil surface during the first growing season.[10] While protection from wind and temperature extremes provided by the nurse crop increases perennial grass or forb germination and establishment, competition for moisture reduces plant growth. During the second growing season, however, when the companion crop does not regenerate or is present in reduced numbers, perennial plant growth is released from this competition. Growth of perennial species is enhanced in the second and subsequent growing seasons because the protection of the nurse crop made possible better establishment during the first growing season.

As a general rule, nurse crops should not be used where annual precipitation is less than 30 cm (12 in.). In low precipitation areas, slower growing perennials suffer a loss of numbers and vigor in the first growing season due to the moisture deficit created by the companion crop.

Temporary stabilizing species should be short-lived annuals or biennials that are adapted to the soils and climate of the region. They should germinate and establish rapidly and have vigorous roots and shoots. While they should not develop from seed during the second growing season, persistent dead plant parts are beneficial. During the second growing season dead plant stems protect the soil surface from the erosive force of high winds and lower the surface soil temperature. Standing litter also holds snow on site and provides shade for the developing plant community. Cereal grains and seed sterile hybrids such as Regreen are commonly used as nurse crops in the northern Great Plains.[11]

SEASON OF SEEDING

Numerous manuals and texts state that seeding should take place immediately prior to the period of maximum precipitation.[12] In the northern Great Plains this season is early spring. In other places, it may be summer or fall. Spring and fall seedings are most commonly used. Both seasons have advantages and disadvantages.

Late fall seedings, which are common on disturbed sites, are referred to as dormant fall seedings. Seed is placed in the ground as late in the season as possible. It undergoes vernalization in the ground and is ready to germinate with the first precipitation of the spring season. Use of equipment and manpower is more efficient since they can be mobilized when convenient and not pulled from other tasks at inopportune times. The disadvantages of fall seedings are (1) the loss of seed to desiccation and depredation by birds and rodents during the long dormant period, and (2) the germination of weeds with the seeded vegetation in the spring.

Spring seedings avoid the vicissitudes of winter and the seed may be vernalized in storage, not in the soil. Cultivation is usually required before spring seeding to destroy any weeds that have germinated. Because of the cultivation, spring-seeded beds have fewer weeds and are usually more productive than fall seedings during the first few growing seasons. Spring seedings appear to have more advantages than fall seeding.[13]

The major disadvantage of seeding in the spring is the difficulty in predicting precipitation events. Because the weather is so unpredictable in semiarid areas, scheduling of equipment is difficult and machinery may not be available when needed. Seeding may be postponed because of scheduling problems or inclement weather. If the seed is not in the ground when it rains it may be impossible to get equipment into the field until the soil dries. This may be after the spring rains — in which case the impetus for germination and establishment provided by spring precipitation is lost. The seeding may be a failure under these conditions or, at best, the seed will not germinate until the next wet season.

As a general rule, cool-season grasses and forbs have their most rapid growth during the cool spring months. Seed must be in the ground to utilize moisture available at that time. Warm-season grasses and forbs grow most rapidly during the warmer summer months and may be seeded later in the growing season if moisture is available. Most shrubs and trees require moisture throughout their spring-summer growing season. They are much harder than grasses to establish from seed on semiarid disturbances.

PLANTING

The transplanting of whole live plants or plant parts into a disturbed soil often is the only viable method of establishing certain plant species. Seeds of some long-lived shrubs or trees germinate only occasionally, establish very poorly, or reveal slow growth rates in the natural environment. It is the rare plant of such species that is able to grow to a mature tree or shrub. For these species, transplanting whole plants or plant parts bypasses the critical seed germination and plant establishment stages. Transplanting ensures that these species have a better probability of producing a mature plant than seeding, but costs are high. The practice requires use of relatively expensive nursery stock and it is labor intensive. Widespread use of transplanting is not recommended, but for special use areas or selected species it may be the only alternative.

Planting may take place prior to or after seeding. If transplanting into an established plant community or into a new seed bed, living plants or seeds of more competitive species should be removed from the area directly adjacent to the transplant. This scalping removes direct root competition from the transplant and improves its probability of successful establishment and growth.[14] Scalping may consist of physical or chemical (herbicide) removal of competing vegetation.

Whole Plants

Whole plants or plant parts may be transplanted. Whole plants may be referred to as bareroot plants, container-grown stock, or wildings. Bareroot plants are shrubs or trees without leaves and without small roots. Container-grown plants are forbs, shrubs, or trees actively growing in some sort of container. Wildings are native plants growing in the immediate area in which they are to be transplanted. They may be grasses, forbs, shrubs, or trees.

Bareroot stock are grown in a protected or enclosed area. When the plants reach a predetermined size, they are induced to reduce their physiological activity, or hardened. During the hardening process, plant physiological activity is inhibited by exposing the plant to reduced moisture, poor nutrient supply, low temperatures, decreased day length, and increased wind. Or simply stated, hardening is the process of inducing dormancy in glasshouse-grown plants by exposing them to ambient conditions they are likely to experience at the planting site immediately after transplanting. During the hardening period (weeks to months), plants increase their carbohydrate reserves and reduce their physiological activity.[15] Hardened or dormant plants survive transplanting better than actively growing plants and are less likely to succumb to the rigors of handling during the transplanting process. Larger carbohydrate reserves further increase the probability that the young plant will survive transplant shock.

Once plants destined to be transplanted become inactive, they may be removed from the root media, stems and roots trimmed, and plants packaged. They must now be maintained in a cool, dark, moist environment until transplanted in the field. Premature removal from the protection of a cool, dark, and moist environment is the major cause of loss of this type of nursery stock. If the plants are permitted to warm up in the glasshouse or in the field, they become physiologically active. If this occurs before they are placed in the ground large losses occur. A recent report stated that 90% of Scotch pine bareroot seedlings succumbed after only a 10 min exposure on a clear hot day.[16] Bareroot stock must be protected at all times — from the moment of shipment from the nursery until it is placed in the ground at the disturbed site.

Containerized stock, including tubelings, are grown in a protected area (glasshouse) in some sort of root container. These plant holders range from individual pots or tubes (Figure 35) to multiple unit packets

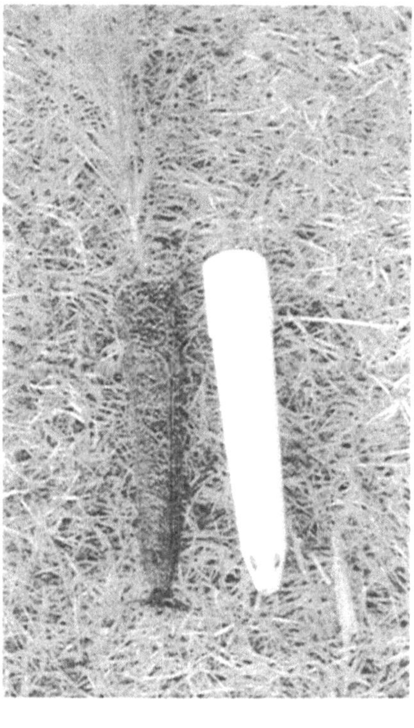

Figure 35 An example of containerized stock. (Picture courtesy of Bitterroot Native Growers.)

(from 6 or 8 to over 100 individual plants in a packet). Juniper bushes (*Juniperus* spp.) with root balls wrapped in burlap or in large buckets, and garden flowers in small 5-cm (2-in.) paper or plastic pots, are examples of common container-grown plants. These plants were seeded or transplanted into packages containing rooting material and nutrients. The package may or may not be placed into the soil when the containerized plant is transplanted. Paper or other biodegradable containers (e.g., peat, cardboard, and cloth) are usually placed in the soil, whereas metallic or other nonbiodegradable containers are removed.

Plants in containers are encouraged to grow very rapidly in the greenhouse. Their growth is usually thin and stringy. But, root hairs and growing tips of roots are protected by this planting technique. The roots are also well developed and firmly established in the rooting medium prior to being placed in the soil. These plants, therefore, are less likely to be injured prior to transplanting or to suffer transplant shock immediately after planting. Growth in a container ensures faster root development and larger and healthier plants after one or two growing seasons than comparably sized bareroot stock or wildings.

Container-grown plants may be purchased in almost any size or shape of package. The only generalization about them is that the entire root mass is included in the plastic, fiber, or paper container. Long, narrow containers (tubes with lengths over 0.5 m [1.5 ft]) encourage the development of deep root systems and have shown promise in semiarid lands. Trees growing in small, 15-cm (6-in.) paper containers in packets of 40 to 100 are often used to rapidly rehabilitate clear cuts or areas subjected to severe forest fires. This type of stock has been very successful when correctly planted.

Wildings are dug from their natural setting and transplanted to the degraded site. Both trees and shrubs have been successfully moved by this technique. Trimming of all wildings is essential to keep plants viable after transplanting. Water loss through transpiration and respiration may be significantly reduced by proper pruning. Most vegetation specialists hesitate to prune adequately and risk loss of the transplants due to excess water loss after transplanting.

The mechanized tree spade is widely used to transplant wildings — either trees or clumps of shrubs. This device may be first used to open a receiving hole, then to dig up the plant selected for transplanting, and finally to place the plant with an accompanying soil and root mass into the previously opened hole (Figure 36). The compacted soil along the sides of the receiving hole poses a penetration problem for the developing shrub or tree roots. This impenetrable soil zone should be destroyed before the transplant is placed into the receiving hole. This may be accomplished by breaking the compacted

Figure 36 Tree spade with wilding.

layer with an axe or appropriate instrument before placing the wilding into the hole. The use of the mechanical tree spade is very successful, but may cause plant loss when the movement of large individuals is attempted. While severe pruning increases the survival rate of all wilding transplants, the root to shoot (above-ground plant part) ratio should be kept close to or greater than 1:1 to insure maximum survival on disturbed semiarid areas where irrigation is unlikely.

The tree spade is slow and expensive. It is rarely used in large-scale revegetation projects when cost is a major limiting factor. Wildings, as well as bareroot and containerized stock, are expensive and receive attention only for critical or intensive use areas. The major advantage of wildings is that their genetic complement is guaranteed to have been selected to meet the soil and climatic extremes of their region of use.

Several types of tractor-powered mechanical tree or shrub planters are available (Figure 37). These devices are much faster than transplanting by hand, but they plant only in relatively straight lines. Shrub groupings created by these planters do not resemble natural communities, but the savings and speed permitted by these devices allow mining companies to place many additional plants in the ground.

Different tree and shrub species have better or poorer survival rates, depending upon the transplanting technique used. In general, wildings are used infrequently in rehabilitation programs because of their low survival rate without intensive management and care. With the

Figure 37 Mechanized tree or shrub planter. (Picture courtesy of Western Energy Company.)

other two techniques, some shrubs and trees perform better as bareroot plants and others as container-grown stock.[17] Bareroot shrubs and trees are light and easy to handle and may be transplanted in the fall or spring. These plants are usually larger and older than container-grown stock, have tough woody stems and roots,[18] and have the protection afforded by a thick bark. However, bareroot stock pose handling and storage problems. In a warm and breezy environment roots may dry out enough to kill the plant in a very short time if the plants are not completely dormant. Hardening is essential to survival of bareroot transplants. Container-grown plants may be hardened or placed in the ground without hardening. In the latter case, the containerized plants continue growing when they are placed in the soil and develop new roots faster than bareroot stock. However, bareroot plants are usually more mature and more likely to survive the vicissitudes of wind, temperature extremes, insect depredation, and rodent attack after they have been transplanted. Bareroot stock is also cheaper than comparably developed containerized plants. While most large-scale attempts to establish woody plant communities use bareroot stock rather than containerized stock, both transplanting methods are used in land rehabilitation programs. Species should be evaluated at each individual site to determine which technique is appropriate for that particular disturbance and rehabilitation program.

Plant Parts

Cuttings, root pads, and sprigs are plant parts used to establish shrubs or trees. Cuttings consist of woody roots or pieces of stems that include nodes. Stem and root buds develop from meristematic tissue in the root or at the plant nodes and grow into complete plants. It is relatively common to dip cuttings into a plant growth hormone prior to burial to stimulate development of the new plant. Any shrub or small tree that commonly reproduces from rhizomes, root sprouts, or adventitious stems may be readily propagated by cuttings. Willows are commonly propagated for riparian ecosystems by these pieces of stems or roots.[19,20] This technique is relatively inexpensive and adapted to the rapid proliferation of large numbers of trees or shrubs.

Root pads and sprigs are used to establish clumps of shrubs rather than single plants. Prior to planting the receiving site must be prepared. This consists of the creation of a slight depression (15 to 20 cm [6 to 8 in.] deep) in loose surface soil. The size of the depression must be appropriate for the desired use, available plant root crowns, and shrub species. Shrub stands may be of any shape, with diameters measuring from centimeters to meters. The shrubs that are to be transplanted are trimmed to a few centimeters (inches) in height before initiating removal of the root pad or the sprigs. The root pad is created when a frontend loader lifts a pad of dirt containing multiple root segments from an undisturbed site. This pad, which is cut as thick as the receiving depression, is set into the prepared location on the disturbed landscape. It is either anchored with sand or topsoil or simply compacted in place. Sprigging requires the removal and movement of only roots and sprouts from the undisturbed location to a prepared site on the disturbance. Sprigs must be covered with sand or topsoil and lightly compacted. Thicket-forming shrubs which spread by rhizomes, root sprouts, runners, or tip layering (e.g., snowberry [*Symphoricarpus* spp.], some wild roses [*Rosa* spp.], and raspberry [*Rubus* spp.]) lend themselves to sprigging or root pad construction.

Special Planting Techniques

Water deficit is the major cause of transplant loss. To overcome the effects of the lack of water, condensation traps and gel-forming, water-absorbing polymers have been successfully used. Condensation traps (Figure 38 and 39) consist of a plastic shield or cone placed on the ground surrounding the transplant.[21] The afternoon sun causes the soil temperature to rise and moisture to vaporize, but if it vaporizes under the shield it condenses on the cooler plastic and runs down

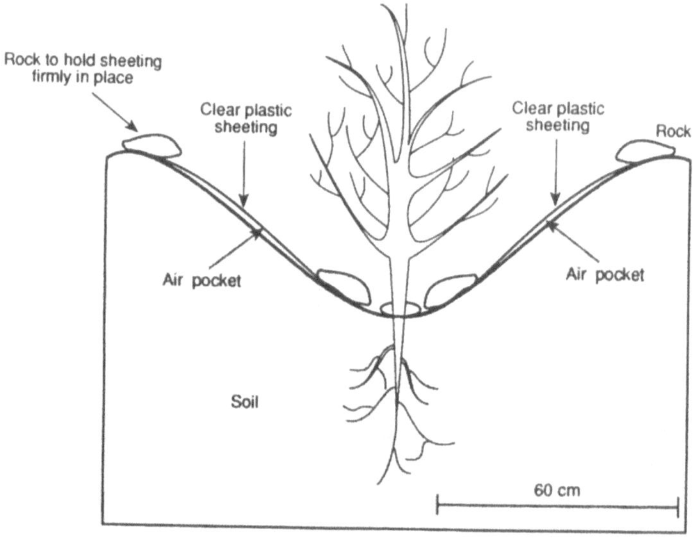

Figure 38 Schematic representation of a shrub in a condensation trap.

Figure 39 Two plants in a condensation trap.

the sheet to the base of the plant, thus irrigating the transplanted shrub or tree. Although installation is labor intensive, this technique improves the survival of dryland plantings of shrubs or trees.

Gels are organic polymers that have enhanced water-holding capacities. They are capable of absorbing and retaining large quantities of water. Small amounts of the polymer are inserted into the hole dug to receive the transplant. Dirt is placed over the polymer and the plant roots, just as if the polymer was not present in the receiving hole. During periodic precipitation, the polymer absorbs and retains water. This water is available to plant roots at some later and presumably dryer time. Planters must exercise care to avoid putting an excessive amount of gel in the receiving hole. Because these gels swell as they absorb water, they may damage roots or push the seedling out of the receiving hole when moisture is abundant.

Sod is also used in disturbed land revegetation. It provides a very rapid means of stabilizing disturbed soils, but sodding, like all transplanting techniques, is expensive. This technology is used only for degraded sites that are subject to severe erosion and must be stabilized immediately. Drainage ways (Figure 40) have been successfully reconstructed by this practice.

Rangeland sod cannot be manipulated like lawn sod. Range plants are rooted deeper, and rhizomes that hold range sod together are located deeper in the profile than rhizomes of Kentucky bluegrass, the main component of lawn sod. Pieces of rangeland sod, therefore, are much heavier than lawn sod — 45-cm (18-in.) squares are about the maximum size of a piece of rangeland sod that can be conveniently handled. The rate of rangeland sodding, therefore, is slow as well as expensive. It is not often attempted on large disturbances.

Figure 40 Sodded highway drainage.

Planting Failure

Water is the limiting factor for almost all properly transplanted shrubs or trees. To insure adequate moisture for plant establishment, transplanting should be carried out during the dormant fall period immediately prior to freeze-up or very early in the spring before the period of maximum precipitation. Transplanting generally takes place in the spring. It is during this season that climatic conditions are optimal for transplant success. Soils are usually moist, temperatures are low but frost heaving should not be a problem, and the probability of major precipitation events is highest at this time in most parts of North America. Nevertheless, even when transplanting takes place during this season and drought is not so severe as to kill the transplant, large numbers of plants die.

Other than drought, the major causes of transplant failure are

1. Lack of adaptation to site characteristics
2. Poor handling immediately prior to planting
3. Poor planting technique
4. Competition from other plants that establish faster than the transplanted species

As with seeded species, whole plants or plant parts selected for transplanting must be adapted to the soils, physical characteristics, and climate of the site. A species is transplanted because it shows growth problems that make seeding a nonviable plant establishment alternative. Seed or plant parts for propagation of nursery stock should be selected from populations in the area to receive the transplants. Collection of seed or native plants at the site of planting ensures that the transplants possess tolerances selected by natural forces acting at the degraded site.

The second obstacle may be at least partially overcome by hardening transplants prior to removal from the protection of the nursery. Plant desiccation in the field (an especially critical factor with bareroot stock and wildings) may be prevented by protecting the plant roots from sun or wind and keeping them cool and moist.

The construction of holding facilities prior to arrival of plants at the disturbance site also helps reduce losses immediately prior to planting. Shipping cartons should be opened upon arrival and water supplied to thoroughly wet the plant root systems. Periodic inspections during holding are necessary to prevent desiccation. Plants should be protected from direct sun, excessive heat, and wind. The dormancy of bareroot stock must be maintained. Not only must these plants be protected from sun and wind, but storage temperatures should be lowered to 1 to 4°C (34 to 39°F). If bareroot stock are exposed to warm temperatures for even a few days before planting, they break dormancy and survival is reduced. When bareroot stock are taken into the field, they are usually in plastic or cardboard containers. These wrappings trap heat and can increase the loss of plants before they are placed in the ground. Even when in the field, bareroot stock must be kept cool.

Poor planting technique is common. It contributes to poor plant survival and growth. Correct technique requires training and supervision of planting crews. Care must be exercised to ensure that roots are placed deep enough into the soil, but not so far as to bury the above-ground part of the plant. No "J" roots should occur at transplanting. Such placement increases loss of plants because root tips are turned up or away from their natural position, which inhibits growth of new roots. Soil around transplants must also be compacted enough to remove air pockets and ensure good soil-root contact, yet not compacted so firmly as to break secondary or tertiary branches of the main root or impede movement of water.

The last problem is less amenable to correction. If shrub or tree planting takes place when forbs and grasses are seeded the transplants usually experience massive dieoffs. The faster growing root systems of the forbs and grasses absorb water and soil nutrients before trees or shrubs can acquire them. The rapidly developing stems and leaves of herbaceous species also shade the transplants. The competition above and below ground kills the new shrub or tree before it can produce root and shoot systems strong enough to compete with other plants. If either containerized or bareroot stock are transplanted into established grass stands they have the same dieoffs. Again, competition for water and nutrients as well as shade appear to be the main causal agents of the plant losses. Transplantings of some species have been carried out before direct seeding, but soil stabilization is inhibited by such programs.

A common practice used to minimize this competition between woody and herbaceous plants is called scalping. This is the creation of a disturbance to remove grasses and forbs from the area around the transplant. This mechanical removal of vegetation[22] or herbicide application to kill plants have both been successfully used in the East and West. While more successful than nontreatment, scalping adds to the cost of tree or shrub transplanting.

REFERENCES

1. Bengson, S. A., ASARCO's revegetation of mill tailings and overburden wastes from open-pit copper mining operations in Arizona, in Proc. 2nd Annu. Meet. Am. Soc. Surf. Min. Reclam., Denver, CO, 1985, 164.
2. Whittaker, R. H., *Communities and Ecosystems*, 2nd ed., Macmillan, New York, 1975, 117.
3. Vallentine, J. F., *Range Developments and Improvements*, 3rd ed., Academic Press, New York, 1989, 270.
4. Young, J. A., Revegetation technology, in *Proc. Symp. Perspectives Processes Rangeland Reveg.*, WRCC-21 and Society for Range Management, Reno, NV, 1990, 92.
5. Cooper, H. W., Some plant materials and improved techniques used in soil and water conservation in the Great Plains, *J. Soil Water Conserv.*, 12, 163, 1956.
6. Seamands, W. J. and Powell, L. M., *Plant Science Fact Sheet*, Wy. Agric. Exp. Sta., University of Wyoming, Laramie, undated.
7. Soil Conservation Service, Plant Materials for Use on Surface-Mined Lands in Arid and Semiarid Regions, EPA-600/7–79–134, SCS-TP-157, U.S. Department of Agriculture, Washington, D.C., 1982, 7.
8. Fulbright, T. E., Redente, E. J., and Hargis, N. E., Growing Colorado Plants from Seed: A State of the Art. Vol. II, Grasses and Grasslike Plants, FWS/OBS-82/29, U.S. Department of the Interior, Washington, D.C., 1982.
9. Tennessee Valley Authority, Aerial Seeding and Fertilization Land Reclam. Prog., Div. of Land and Economic Resour., Norris, TN.
10. Vallentine, J. F., *Range Development and Improvements*, 3rd ed., Academic Press, San Diego, CA, 1989, 280.
11. HybriTech, *A Cool Season, Soil Stabilizing Cover Crop*, Inf. Circ., HybriTech Seed Int. Inc., Witchita, KS, 1991, 1.
12. Wasser, C. H., Ecology and Culture of Selected Species Useful in Revegetating Disturbed Lands in the West, FWS/OBS-82/56, U.S. Department of the Interior, Washington, D.C., 1982, 3.
13. Hart, R. H. and Dean, J. G., Forage establishment: weather effects on stubble vs. fallow and fall vs. spring seeding, *J. Range Manage.*, 39, 228, 1986.
14. Sloan, J. P. and Ryker, R. A., Large Scalps Improve Survival and Growth of Planted Conifers in Central Idaho, Interm. Res. Sta. Res. Pap. INT-366, U.S. Department of Agriculture, Ogden, UT, 1986, 7.
15. Penrose, K. D. and Hansen, D. J., Planting techniques for establishment of container-grown or bareroot plants, in Shrub Establishment on Disturbed Arid and Semiarid Lands, Proc. Symp. WY Game and Fish Dept., Laramie, 1981, 37.
16. Laursen, S. B. and Hunter, H. E., *Windbreaks for Montana*, Coop. Ext. Serv. Bull. 366, Montana State University, Bozeman, 1986, 17.
17. Bjugstad, A. J., Yamamoto, T., and Uresk, D. W., Shrub establishment on coal and bentonite clay mine spoils, in Proc. Symp. Shrub Establishment on Disturbed Arid and Semiarid Lands, WY Game and Fish Dept., Laramie, 1981, 104.

18. Stevens, R., Techniques for planting shrubs on wildland disturbances, in Proc. Symp. Shrub Establishment on Disturbed Arid and Semiarid Lands, WY Game and Fish Dept., Laramie, 1981, 29.
19. Monsen, S. B., Plants for revegetation of riparian sites within the Intermt. Region, in Proc. Symp. Managing Interm. Rangelands — Improvement Range Wildlife Habitats, Gen. Tech. Rep. INT-157, Interm. For. and Range Exp. Sta., U.S. Department of Agriculture, Ogden, UT, 1983, 83.
20. Schultze, R. J. and Wilcox, G. I., Emergency measures for streambank stabilization: an evaluation, in Riparian Ecosystems and Their Management, Reconciling Conflicting Uses, Gen. Tech. Rep. RM-120, Rocky Mount. For. Range Exp. Sta., U.S. Department of Agriculture, Fort Collins, 1985, 59.
21. Jensen, I. B. and Hodder, R. L., Tubelings, Condensation Traps, Mature Tree Transplanting, and Root Sprigging Techniques for Tree and Shrub Establishment in Semiarid Areas, Agric. Exp. Sta. Res. Rep. 141, V. 2, Montana Department of State Lands, Helena, 1979.
22. Sloan, J. P. and Ryker, R. A., Large Scalps Improve Survival and Growth of Planted Conifers in Central Idaho, Interm. Res. Sta. Res. Pap. INT-366, U.S. Department of Agriculture, Ogden, UT, 1986, 7.

5

Special Reclamation Practices to Enhance Vegetation Diversity and Usefulness

The first four chapters of this text are devoted to the principles and practices that contribute to successful revegetation of disturbed lands. While important, simple establishment of a few plant species does not contribute to the full potential of the ecosystem to prevent erosion, support livestock and wildlife, and present an aesthetically pleasing landscape. Multiple species plant groupings meet these goals much better than one or two species seedings. Ecologists also agree that plant diversity or species richness contributes to greater community stability and enables the vegetation to better meet multiple uses. Diversity, therefore, is one of the goals of the revegetation program.

Interspecific and intraspecific competition occur within plant communities. Adjacent seedlings of the same or different species are in competition for water, nutrients, light, space, etc., as soon as the radicle penetrates the seed coat. But different species of grasses, forbs, shrubs, and trees have varying physical, chemical, and biotic needs as a result of differing genetic compositions. Certain plant species coexist in the same habitat because their genetically controlled requirements for germination, establishment, and growth complement one another and they do not directly compete for the same nutrients, water, light, etc. Variations in site characteristics also produce numerous niches, microenvironments, or microhabitats, and different species germinate and establish better in one niche or another. Microvariations include a multitude of factors such as minor fluctuations on the land surface or plant location with respect to a rock, larger plant, snow drifts, sheltered areas, etc. This type of site variation extends across very short distances but creates change or different environmental factors within plant communities. It permits a wide range of genetic expression by numerous species, populations, and individual plants within a species.

Macrohabitat-influencing characteristics include aspect, slope position, elevation, climate, and numerous soil factors. The most influential soil factors are those expressed in the surface or near-surface soil layer. They include factors such as texture, salinity, acidity, organic matter, nutrient levels, etc. All of these factors exert strong influences on plant development, but since the scale is so much larger than the microhabitat variation previously noted, these factors contribute to the creation of different groups of plants over large areas or plant communities.

In the natural environment, plant communities vary across the landscape as any macrohabitat characteristic changes. Reclamation specialists attempt to duplicate the natural environment in their restoration efforts. However, on disturbances from highway shoulders to surface mines, macro- and microenvironmental variables have been destroyed. After recontouring, aspect and slope are much more uniform on the disturbance than in the natural environment. It is not economically feasible to construct the variation found in these parameters on the undisturbed landscape. Changes in soil chemical and physical characteristics are also lost during surface soil stripping and reconstruction. After removal, possible storage, and respreading, topsoil as applied to a degraded site is a more thoroughly mixed and uniform mantle than that which existed on the site prior to disturbance.[1] The component of floral diversity contributed to the original plant community by surface soil variation and different slopes is lost in rehabilitated ecosystems.

Specialists responsible for early revegetation efforts assumed that after seeded vegetation had germinated and established, desirable species from surrounding ecosystems would invade the new stand. This influx of new species would contribute to the diversity of the new plant community and enable the disturbed site to meet bond release criteria. Rehabilitation scientists have found that plant cover has an inhibitory effect on the invasion of a site by new plants. This phenomenon appears to be more strongly expressed as the disturbance becomes more arid. Once grass cover is successfully established on the Great Plains, it tends to resist invasion by new species[2] even though this invasion would increase site stability, augment diversity, and contribute to community succession. Successful development of a stand of vegetation on a degraded site does not stop movement of new species into the community but it does impede the movement. Natural plant invasion of disturbances in semiarid to arid regions occurs in such small numbers that the process cannot be included in those factors that contribute to an increase of diversity on the site during the minimum bonding period. The revegetation process and management become of utmost importance to the development of plant diversity on degraded sites.

The invasion of seeded plant aggregations on degraded sites is slow because seed distribution is restricted and because the germination and seedling establishment phases of plant growth are sensitive to site characteristics such as moisture deficit, light levels, and temperature extremes. New plant recruitment will occur only when seed is present and germination, establishment, and plant growth are successful. The three latter phases of plant recruitment (germination, establishment, and growth) can only take place when plant habitat requirements match those at the particular site in which the seed breaks dormancy. As the environment becomes more arid, this combination of seed presence, species requirements, site characteristics, and environmental conditions occurs with increasing rarity. Seed presence, sufficient moisture, correct temperature, or other factors necessary for seed germination and establishment are not available every year at the proper time and sequence to guarantee good seedling recruitment. On semiarid to arid lands new plant recruitment for at least some species is regulated by episodic events, or a series of episodic events, rather than average environmental conditions.[3] Thus, seed of a particular species may germinate only during years of wet springs, while seedlings may successfully establish only during years with wet summers. Therefore, the species can develop new plants only when viable seed is in the ground and growing season precipitation is above average in both spring and summer. In addition to the limitations imposed by these environmental conditions, competition between the germinating seedlings and other seedlings or established plants will curtail plant recruitment.

Despite these impediments to nonseeded plant establishment, a degraded site will eventually reveal an increasing number of plant species that were not in the original seeding. The time interval for this invasion may be only a few years in the mesic eastern parts of North America; it may require many decades in the more arid Great Plains. The probability is low that a series of events will terminate in the successful recruitment of an invading species on a degraded site that had been previously revegetated. Therefore, if restoration is to be successful in a reasonable number of years, the revegetation specialist must adopt means other than simply waiting for natural invasion to diversify the plant communities on semiarid to arid disturbances.

The most obvious first step to encourage the generation of diverse communities across the disturbance is to reconstruct the physical environment for each community (i.e., soil, exposure, elevation, etc.) and to seed distinctly different seed mixes in each different habitat. Those responsible for restoring vegetation on disturbances rarely find this an inviting alternative. The expense and effort required (1) to develop several unique seed mixes; (2) to locate, delineate, and construct the various habitats; and (3) to change seeding techniques and mixtures to match the different plant aggregations, are greater than the expense and effort of activities presently undertaken to meet the goal of plant diversity on disturbed lands.

The reclamation industry has utilized broad seed mixes of numerous grasses, forbs, and shrubs to establish diverse plant communities. This has provided a limited amount of diversity on some sites, but it has not been universally successful. To meet the legislated goal of diversity on coal-mined lands, the reclamation industry is trying new techniques and innovative approaches to total revegetation.

Some of the procedures developed by coal mine revegetation specialists, and to be discussed in this chapter, are already in practice, but some are in a trial or research stage and do not meet the requirements of reclamation laws or guidelines of federal or state agencies. While all of these practices contribute to landscape or community diversity, they are not universally accepted as part of reclamation efforts by all segments of the revegetation industry. They deserve mention, however, because they contribute to complete restoration and the aesthetic appearance of disturbed landscapes. Their future acceptance in rehabilitation programs will be determined by their contribution to the structure of plant communities and stability of disturbed landscapes as viewed by regulators, revegetation experts, and the general public. This chapter will discuss several of these methods of increasing diversity on disturbed lands.

COMMUNITY DIVERSITY

Many botanists find it hard to accept that the number of species that grow in the natural, undisturbed environment are expressing unique niche requirements.[4] However, when the number of semiarid grassland species that develop significant cover within one community (usually 8 to 16 species reveal cover greater than 1% in a Northern Great Plains plant community)[5] is compared to the possible variation in germination and establishment characteristics, it becomes entirely possible, if not probable, that each species is exploiting a different niche. To produce a similar range of plants on a degraded site, the revegetation specialist must seed a comparable number of compatible species and reproduce a range of niches equivalent to that found in the original ecosystem. While the selection of a number of compatable species is possible, the creation of the many microenvironments on a disturbance is an almost impossible task. However, a number of alternatives are available to the revegetation specialist to help increase species richness within a plant community on a disturbed site.

The alternatives that enable revegetation experts to increase the number of plant species in disturbed land communities require departure from or addition to accepted revegetation techniques. Several basic methods of fostering either diversity within plant communities or entirely different communities on degraded landscapes are

1. Seeding broad seed mixtures of numerous well-adapted species
2. Utilizing different seed mixes in different macrohabitats
3. Seeding reduced seed rates for aggressive and competitive species and increased seed rates for less competitive plant species
4. Applying direct-haul topsoil
5. Spreading native hay mulches instead of mulches composed of stems of cereal grains
6. Planting islands of native sod within the disturbance
7. Supplemental interseeding or planting
8. Seeding on multiple dates
9. Management practices that increase diversity

Some combination of these activities will be necessary to achieve the level of diversity required by modern reclamation laws or restoration activities.

Revegetation specialists have relied almost exclusively on two activities to meet their botanical diversity goal. As previously noted, the most common activity practiced on degraded sites to stimulate the development of multiple-species plant communities has been the use of broad seed mixes of numerous grasses, forbs, and shrubs. The second technique, now gaining wider acceptance, is the seeding of different combinations of a few basic seed mixes. For example, at the Western Energy Coal Mine in southeastern Montana, a basic upland mix of six grasses and one shrub is drilled over most of the disturbed area. This seeding is followed by broadcasting a supplemental mix of three warm-season grasses and five forbs. A different lowland mix is drilled in annual flood plains along drainage bottoms. On ridges or other sites where the company anticipates pine tree planting, a conifer mix is seeded prior to planting of tubelings of Ponderosa pine and skunkbush sumac. On landscapes designated as shrub/grasslands, sagebrush (*Artemisia* spp.) is interplanted into the newly seeded community.[6] The use of these multiple seed mixes has increased diversity on mined lands, but it is yet to be determined if the communities resulting from these seedings will fulfill the requirements of coal mine reclamation laws and regulations.

Disturbances today are often large (many hectares). Multiple unique seed mixtures are essential to the rapid establishment of diverse plant assemblages in different habitats of such large areas. In general, this practice has been resisted by revegetation specialists because of its expense. One broad seed mix has been substituted for several unique seed mixes, with the anticipation that different species will establish in different habitats and invasion by nonseeded plant species will increase diversity. The probability of seed being carried to a particular spot on a degraded soil where the species micro- and macroenvironmental requirements will be met is reduced as the size of the disturbance[7] and the distance downwind from a native seed source increase.[8] Invasion of large disturbances in semiarid regions by large numbers of desirable species within a reasonable amount of time will not occur.

Not only will seeded species inhibit invasion and succession on disturbed lands, but the rapid establishment of more aggressive species in a seeded area will suppress the establishment of less competitive species. To stimulate the development of the slower developing shrubs, forbs, and warm-season grasses, the seed rates for these species should be increased and seed rates for aggressive species reduced. While this may produce strange plant aggregations during the first few growing seasons, as aggressive species complete their development the community will assume an appearance more characteristic of vegetation in the seeded area.

Direct-haul topsoil (Figure 41) is one of the better means of encouraging the long-term development of diverse native plant communities on disturbed lands. This type of topsoil contains viable plant propagules and the microflora necessary for the successful establishment of many native plant species.[9,10] During the period from topsoil application to seeded plant establishment, plant propagules carried to the site in the surface soil layer begin development and growth.

Figure 41 Plant propagules in direct-haul topsoil.

These plants may become established before the seeded species dominate the site.

While the growth of nonseeded native plant species will be inhibited by seeded species, the presence of native plant seeds and plant parts in the topsoil will permit the germination and/or development of nonseeded species when the opportunity presents itself. That is, when something causes the death of a seeded species and opens the vegetation canopy for another plant, sites covered with direct-hauled topsoil are more likely to respond with a native species replacement than are nontopsoiled sites or disturbances covered with stored topsoil or substitute materials.

Use of native hay mulches present another valuable way to enhance the development of nonseeded species on degraded lands.[11] Native hay can be cut and bailed when seeds of the desired plant species are mature. By manipulating cutting dates the reclamation supervisor can change the composition of the seed bank in the hay and can acquire maximum numbers of those seeds most valuable to the creation of the plant community planned for the disturbance. Mulching with such selectively prepared hay can make a significant contribution to diversity in the plant community on the disturbed soil.

Islands or plugs of native sod can also be placed within the seeded plant community to introduce an adapted native seed source into the degraded area. These small sources of native seeds become more important as the size of the disturbance and the distance from a native seed source increases. Such islands may be transferred from native range or shrub lands to the disturbed site with front end loaders, vermeer tree spades, or any other convenient piece of equipment. The effects of these islands of native seed will be long term, and an immediate increase in the number of nonseeded species resulting from this practice should not be anticipated.

Sodding is also a method of encouraging diversity. This practice is expensive but yields good results, both as a means of inserting a native seed source into the rehabilitated area to stimulate succession and as a means of stabilizing critical erosion-prone areas. Native range sod should be cut thicker than lawn sod because it contains plants with deeper and more diffuse root systems. To keep range sod alive after cutting and during transportation, it must be cut several centimeters thick.[12] Because the sod is very heavy, it must be handled in blocks rather than the typical rolls of lawn sod. To maximize the probability of success, native rangeland sod should be transplanted in the spring and watered immediately after transplanting.

Interseeding or planting of specific species that are not adequately represented in the plant community has been attempted to increase diversity. Well-established vegetation reduces the probability of successful interseeding; therefore, it is only carried out in poorly established plant communities or on degraded rangeland. To transplant vegetation into an established community, it is necessary to create an opening for the new plant. Physical or chemical (herbicide) scalping has been used successfully to create openings and reduce competition for transplants or interseeded species in an established community.[13]

The use of multiple seeding dates encourages the survival of less aggressive forbs, shrubs, and warm-season grasses in mixed plant communities.[14] It has been proposed that seedlings of less competitive life forms show better survival on violently disturbed soils when they are not in competition for moisture and plant nutrients with faster growing cool-season native grasses or agronomic species. The first seeding or phase, of a two-phase seeding typically consists of a broadcast seeding of forbs, shrubs, and warm-season grasses. Cool-season grass species may be interseeded before any subsequent growing season. Unfortunately, large amounts of bare ground remain even after establishment of the first phase of a multiple phase seeding. Weed infestations degrade these seedings. Downy brome is a common invader of plant communities when this method is used on the northern Great Plains.

Although little research directly addresses the impact of management of revegetated lands as a tool to increase diversity, much observation and experience supports this concept. For example, mines in southeastern Montana have successfully used fire to control downy brome infestations and to stimulate the growth of more diverse plant communities on minesoils. Several coal mining companies have attempted to stimulate less common species by manipulating livestock grazing to remove or lower litter accumulations and open the grass canopy to other plant species. Mowing or heavy grazing early in the growing season has also reduced the competition presented to forbs and warm-season grasses by faster growing introduced grasses.[15]

The revegetation specialist must utilize combinations of surface soil manipulation, seeding and planting techniques, and management to foster diversity within seeded plant communities. Other programs may be incorporated with these activities to accentuate change in vegetation across the disturbed landscape. Some of these other programs will be reviewed in the next section.

LANDSCAPE DIVERSITY

The change in plant communities across a landscape or the turnover of species across space is a component of the total diversity of the study area. Most natural areas contain multiple plant communities with their accompanying ecotones between the communities (Figure 42), but even large disturbances (e.g., surface mines, tailings disposal

Figure 42 The change of plant communities across a natural landscape in western North America.

areas, and abandoned mined lands) are revegetated with little or no change in plant communities across the landscape. Because of the expense required to seed different combinations of plant species, one or a limited number of seed mixes are drilled or broadcast in most degraded soils. The single seeding usually contains a broad mixture of species, including a number of grasses, forbs, and possibly some shrubs. It is assumed that different combinations of seeds will successfully establish in different micro- and macrohabitats. However, few micro- or macrohabitat-influencing factors remain on disturbed lands. Seeded plant germination, establishment, and growth will be determined by the overriding influence of climate and topsoil. Therefore, while a broad seed mix may be utilized, the species that germinate and establish in one particular growing season will be relatively uniform throughout the seeded area. If the same seed mix is utilized on a different site or in another growing season the vegetation response may be entirely different because climatic patterns and surface soil coverings are rarely identical from one year to the next.

There are, however, several activities that can contribute to the development of landscape diversity at large disturbances such as strip-mines. These may be completed at either a savings or a limited increase in expense to the revegetation program. None of these techniques is widely accepted by both the industry creating the disturbance and the regulatory agency controlling revegetation. In some instances the technique is thought to be expensive (e.g., variable depth topsoiling), in others the activity is thought to represent an exception to reclamation laws or regulations (e.g., bluff extensions). Nevertheless, when properly applied all of the activities mentioned in this section contribute to landscape diversity and enhance the potential for multiple use. Aesthetic appeal of large disturbances will also be improved by increased landscape diversity.

Variable Depth Topsoiling

In the natural undisturbed environment the macrohabitat characteristic surface soil depth varies from maximum thickness in lowland woody draws and drainages to minimum amounts or no topsoil on wind-swept ridges (Figure 43). Plant species, community composition, and plant production reflect the thickness of this soil layer as well as other environmental factors operative in the different vegetation zones.

Topsoil depth never varies on the postdisturbance landscape. It is applied to the recontoured disturbance in a uniform depth unlike that found in the predisturbance environment. As with the loss of different slopes, aspects, and chemical and physical properties, the loss of variation in topsoil depth reduces the inherent variability of the rehabilitated ecosystem. The large number of species found in natural

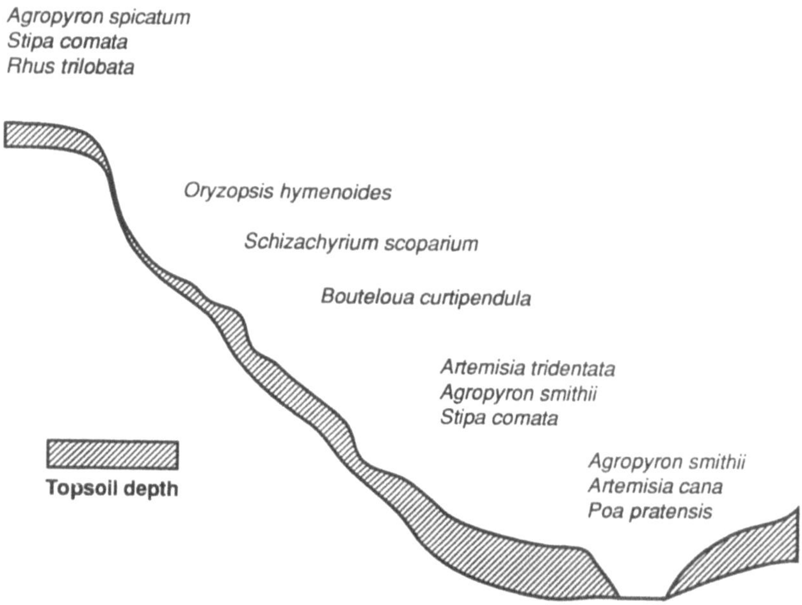

Figure 43 Topsoil depth variation and vegetation response on a natural hill slope in an arid or semiarid area.

ecosystems cannot, therefore, be duplicated in the reclaimed environment. Regulatory emphases on uniform topsoil depth and plant production limit the ability of planners to incorporate variable topsoil depths into reclamation programs and hence impede the development of diverse plant communities across the rehabilitated landscape. The incorporation of varying depths of topsoil from top to bottom of slopes will require a waiver of federal and/or state regulations at most disturbances. However, without the waivers it will be impossible for reclamation specialists to develop the plant diversity across the landscape considered so essential by legislation such as the Federal Surface Mining Control and Reclamation Act.

Bluff Extensions

In the western coal fields, when surface mining is terminated by steep slopes that make stripping ratios prohibitively expensive, so-called bluff extensions are becoming an increasingly popular rehabilitation tool. A bluff extension is an exception to the law requiring reduction of highwalls to 3:1 slopes. In the application of this practice, existing cliffs or ridges are extended to some point between the premine soil elevation and the bottom of the pit. While highwalls should be reduced as nearly as possible to their original contour, the judicious use of bluff extensions can contribute much needed landscape diversity and unique wildlife habitat to disturbed areas in semiarid to arid areas. Bluff extensions can be permitted only when the lithology of the highwall indicates that it will resist erosion and when precipitation is low enough to protect the integrity of the slope materials. The use of bluff extensions in more mesic areas is less clear because of the probability of instability caused by greater precipitation.

Bluff extensions offer several unique wildlife features to the postmine landscape. For example, at several mines in the Powder River Coal Basin, holes in extended highwalls have been excavated with explosives. These openings provide nesting sites for the rare prairie falcon and other birds of prey. Bluff extensions may also serve as sites for the growth of unique wildlife food plants. Steep slopes in the western plains often support shrubs that are utilized by deer or antelope when climatic conditions are most severe. Only under the most extreme snow conditions will animals venture out on these slopes to crop the shrubs, and only then because all other food is inaccessible under the snow. Many of these plants are at the limit of their range and cannot support persistent grazing or compete with better adapted species on more fertile or mesic sites. On very steep slopes, these shrubs are protected from livestock and frequent wildlife browsing by the reluctance of livestock[16] and, presumably, wildlife to feed on exposed steep slopes.

Use of standard rehabilitation equipment or practices to revegetate these areas is not possible. The steep slopes cannot support a uniform cover of grasses, forbs, or shrubs because of aridity, temperature extremes, and lack of a soil covering. In the natural landscape these slopes are havens for scattered shrubs, a few unique forbs, and an occasional clump of drought-tolerant bunch grasses. After disturbance, bluff extensions should be revegetated with these same species. At the vast majority of sites, this will mean shrub plantings or seedings. Appropriate species for seeding or planting will vary by latitude, elevation, exposure, and precipitation.

Characterization of the thin soil mantle and species of wildlife utilizing the area will directly influence vegetation selection. The best guide for selection of appropriate plant species is the premine vegetation inventory. Seed for bunch grasses found on these slopes are usually available from commercial sources, but the forb and shrub component of the communities are unique and not often found in seed catalogues. Seed should be collected locally and either sown directly, grown in containers, or raised in nurseries for bareroot stock. Varieties of the same or closely related species from sites several hundred miles distant should not be used in these seedings. The survival of any vegetation on bluff extensions is precarious. Seeding or planting should utilize populations naturally selected for the soils and habitat of the site of the disturbance.

Impoundments

Growing recreational demands on reclaimed landscapes, public demands for water, and decreasing number of western potholes have made mined land impoundments increasingly valuable. Nevertheless, federal and state regulatory agencies in the west hesitated to permit impoundments on coal mined lands for many years. These water bodies did not meet the requirements of AOC (approximate original contour). It was assumed that water quality would not meet livestock or wildlife use standards. Such innovations on disturbed lands, therefore, were considered out of compliance with regulations.

In the 1970s, the Peabody Coal Company was granted a mine permit which included a provision to develop an experimental final pit impoundment at the Big Sky Mine in southeastern Montana. By 1980, the water budget indicated that the impoundment was stable at an approximate maximum depth of 7.5 m (25 ft) and water quality was better than that found in most wells and stock water ponds in the area.[17] Both wildlife and livestock use of the pond was extensive. Since this demonstration, and because of growing industry and regulatory awareness of the value of these water bodies, regulatory agencies in the northern Great Plains have significantly increased the number of impoundments permitted on mined lands.

Impoundments on coal and bentonite mined lands in the West are formed in depressions created by the removal of soil, overburden, and minerals (Figure 44). Only rarely are permanent dams constructed to pond water. Mined land impoundments are depressions that collect runoff and/or ground water from the surrounding watershed and temporarily retain it on the site. They slowly release this moisture to the ground or surface water system of the area. The increased available moisture in the immediate vicinity of the impoundment encourages the development of different plant communities across an otherwise uniform terrestrial landscape (Figure 45). Water bodies provide a livestock and wildlife water source especially valuable in arid and semiarid areas. Impoundments are also becoming increasingly valuable as nesting habitat for waterfowl as prairie potholes and wetlands continue to be drained for agricultural production or housing projects.

Impoundments on stripmines fall into two categories: final pit impoundments and internal ponds. If the final cut in a stripmine is not backfilled, if it is below the premine watertable, if it receives an adequate inflow of ground and/or surface water, and if the depression will hold water, a final pit impoundment may be developed. If ponds internal to the final cut of a stripmine or if depressions created by open-cut mines meet similar criteria, they may also be developed into year-round water sources.

Figure 44 Impoundment on a western surface mine.

Figure 45 Vegetation influenced by impounded water on a semiarid disturbed landscape.

Open-pit mines, either metal or other minerals (e.g., gold, talc, and bentonite), are rarely controlled as closely as coal mines, so water bodies are often the proposed common fate of these disturbances. A few large open-pit mines have been reclaimed to impoundments in this country and these water bodies have made valuable contributions to the local landscape. The Illinois Coal Basin contains numerous examples of abandoned mine pits that have revegetated to plant-lined open water bodies that are invaluable fish and wildlife resources. Ponds and lakes are assets, and their use in rehabilitation programs should be encouraged.

When impoundments are permitted, slopes must be shaped to guarantee safe access by man, livestock, and wildlife. Established vegetation should stabilize the slopes and shoreline. If short segments of the shore of an impoundment will not be accessible to livestock and wildlife because of the angle of the final highwall, such slopes do not necessarily require reworking (Figure 46). Extensive reduction of highwalls around final pit impoundments will only reduce the storage capacity of the reservoir and extend the area of disturbance into otherwise undisturbed terrain. If the major portion of the shoreline is accessible to wildlife and livestock and the extent of steep slopes is limited, the slopes should be retained to increase habitat diversity and limit the growth of the area of disturbance.

The question of water quality is critical in any discussion of water bodies on degraded lands. If water quality cannot meet the stated

Figure 46 Final pit impoundment with intact highwall.

long-term goals of the impoundment, the water body cannot be permitted. Studies and modelling will only suggest what chemical characteristics the pond will have after reclamation. If mining companies wish to incorporate impoundments into stripmine or open-pit mine reclamation plans, they must be prepared to drain and fill the impoundment if the quality of the water does not meet standards of proposed use.

Reduced sulfur minerals in the ore at some mines pose the problem that oxidation of these compounds will generate acid. The resulting acid minesoils and drainage are a major problem facing rehabilitation scientists at such locations today. The rapid rise of toxic water in the Butte, Montana open pit after it was closed in 1980 serves as a grim reminder of what can occur at these disturbances. Lakes or ponds contaminated by acid are not acceptable rehabilitation products.

Wetlands occasionally develop on rehabilitated disturbances because of altered ground or surface water flows. The frequency of their occurrence can be increased by planning for them during the permitting process. The construction of wetlands is now an accepted activity on abandoned mine lands. They are used to treat acid mine drainage and to improve the quality of water leaving degraded lands.[18] Natural or constructed wetlands on any disturbance make a unique and valuable contribution to plant diversity in a relatively unchanging terrestrial ecosystem. Unfortunately, the addition of impoundments and wetlands to disturbed lands, especially lands disturbed for the extraction of coal, is restricted by state and federal law unless wetlands existed on the site prior to mining.

The remarkable contribution of aquatic habitat features to an otherwise uniform disturbed land ecosystem makes impoundments and wetlands desirable additions to any degraded landscape. Widespread development of these mined land features should be encouraged where possible under existing site conditions of rainfall and ground water flows.

A rarely implemented but valuable modification of the highwall extension concept incorporates a final pit impoundment at the base of a bluff extension. This technique requires the waiver of the requirement to return the site to AOC in addition to the waiver of the requirement to backfill the final pit. Release from two regulatory requirements is almost unknown in U.S. coal fields despite the obvious benefits of this landscape feature to wildlife and livestock.

Revegetation in and around impoundments and wetlands requires completely different seed mixes from those applied to the major portion of the disturbed site. Vegetation around the pond will have access to supplemental water throughout most of the growing season. Species selection should emphasize this extra water supply, and plant species better adapted to utilize this moisture should be seeded around impoundments and wetlands. These plants are more productive than dryland species and total forage production around water bodies will reflect both species selection and increased moisture availability. Since wildlife use is one of the major justifications for the placement of impoundments on disturbed lands, vegetation plantings and seedings adjacent to these sites should meet food and cover needs of the wildlife expected to return to the area. Riparian vegetation including trees, shrubs, forbs, and grasses should be seeded or planted adjacent to the anticipated high water line. Aquatic vegetation should be planted below the minimum water level. Fish and aquatic invertebrates should be placed in the water body as soon as possible after accumulation of adequate water to support the organisms. The state Fish and Game Department should be consulted for appropriate plant and animal species for propagation in impoundments.

Impoundments on reclaimed landscapes usually become loafing areas for cattle or horses. Vegetation around the pond is destroyed and the land becomes a sacrifice area to livestock use. To protect these areas from abuse a major portion of the shoreline should be fenced. This will prevent livestock destruction of the vegetation and encourage wildlife use. These protected areas will become wildlife reservoirs on the disturbed landscape.

Inplace Spoil Ridges

There is no doubt that spoil piles at almost all stripmines should be reduced to AOC, as required by federal and state regulations.

However, many western mines are disturbing areas occupied by landforms referred to as breaks or gumbo knobs. Before mining, these lands are sparsely vegetated with a few hardy grasses, forbs, and shrubs. An occasional clump of small trees may be found in protected areas where soil fines have collected and salts have been leached or flushed from the soil. Although this habitat generally cannot support crop production and has little livestock grazing value, these areas are often critical to several classes of wildlife during winter months. After rehabilitation, the wildlife food and cover resources that made these areas unique are replaced by livestock pastures. At least one example of the retention of short stretches of spoil ridges to replace rough-breaks habitat has been developed at the Western Energy Mine on the northern Great Plains.

Construction of this type of topography on degraded sites requires certain restraints. The areas should not be topsoiled and the spoil material used for their construction must be nontoxic and of a texture that resists erosion. Clays or silty clay spoils, rather than coarse-textured materials, will probably provide a more erosion resistant substrate. It may be necessary to strike off the ridge tops and cap the spoil with a consolidated erosion resistant material to prevent excessive soil movement.

Because these hills do not support extensive grasslands, of necessity, vegetation will be sparsely seeded and planted. Species selection for these sites is as critical as it is for bluff extensions. Predisturbance presence on similar topographic material is the most appropriate determinant for revegetation selection. Extensive transplanting of bare-root or containerized stock grown from seed collected on the site prior to disturbance may be necessary to achieve the development of wildlife food plants that are so important to the complete functioning of this ecosystem. In some cases, even seed to be disseminated on the slopes of the spoil ridge should be collected on-site prior to disturbance.

This type of reclamation may not be appropriate in mesic areas, but it may be incorporated into arid or semiarid land rehabilitation programs on an experimental basis. If incorporated into programs with impoundments, such areas will provide unknown wildlife-enhancing potential.

Rock Ledges

Many coal stripmines in the semiarid and arid West are located in areas of irregular landforms that are characterized by rock-capped ridges or exposed rock ledges. At these mines the topography of final slopes may be broken by construction of rock outcrops (Figure 47). These rock structures should be composed of native materials pushed

Figure 47 Rock ledge constructed on a coal surface mine.

together with heavy equipment. Their function is to enhance aesthetic appeal, provide topographic relief, and encourage return of small animals to the reclaimed site. Track and scat counts around several examples of these structures in the Powder River Coal Basin in Wyoming indicate that rock piles are very successful. If large enough and properly integrated with the surrounding ecosystem, small-mammal populations will proliferate immediately adjacent to the rock piles. These animals serve as a food source for other animals located higher up the food chain. Properly designed rock outcrops become the basis for complex food chains as well as a means of visually integrating the degraded landscape into surrounding areas.

Most of these structures are simply piles of rock, but they may be more carefully constructed and incorporated into hillslopes to resemble native rock ledges. Such an application of this technique took place at the Black Thunder Mine in Wyoming. At this mine, walls of rock have been carefully placed on the contour on unfinished slopes. Subsoil and topsoil were placed over these ledges but the face of the rock was left exposed. These rock faces closely resemble native rock ledges.

As with other structures constructed on minesoils, vegetation seeding should reflect the nature of the seedbed. Seeding of native species found on or adjacent to these structures in native habitat should be carried out as soon as possible after construction. These rocky areas often provide a site for weed growth and seed dissemination if they are not immediately planted to more desirable vegetation species.

REFERENCES

1. Schafer, W. M., Variability of minesoils and natural soils in southeastern Montana, *Soil Sci. Soc. Am. J.*, 43, 1207, 1979.
2. Hull, A. C. and Klomp, G. J., Longevity of crested wheatgrass in the sagebrush-grass type of southern Idaho, *J. Range Manage.*, 19, 5, 1966.
3. Call, C. A. and Roundy, B. A., New perspectives in rangeland revegetation, in *Proc. Symp. Perspectives Processes in Rangeland Reveg.*, Meet. Soc. Range Manage., Reno, NV, 1990, 1.
4. Grubb, P. J., The maintenance of species-richness in plant communities: the importance of the regeneration niche, *Biol. Rev.*, 52, 107, 1977.
5. Munshower, F. F., Neuman, D. R., and Rennick, R. B., *The Effects of Stack Emissions on the Range Resource in the Vicinity of Colstrip, Montana*, Reclam. Res. Unit Publ. No. 8703, Montana State University, Bozeman, 1987, 131.
6. Martin, P., personal communication, Western Energy Co., Colstrip, MT.
7. Lovell, D. S. and Munshower, F. F., Nonseeded species invasion of 12 revegetated surface mined sites at Colstrip, Montana, in *Proc. Billings Mined Land Reclam. Symp.*, Reclam. Res. Unit Publ. No. 9301, Montana State University, Bozeman, 1993.
8. Van Zalingen, S. F., Factors Affecting the Native Species Invasion of a Reclaimed Subalpine Minesoil near Grande Cache, Alberta, M.S. thesis, Montana State University, Bozeman, 1983, 65.
9. Iverson, L. R. and Wali, M. K., Buried, viable seeds and their relation to revegetation after surface mining, *J. Range Manage.*, 35, 648, 1982.
10. Ross, D. J. and Cairns, A., Nitrogen availability and microbial biomass in stockpiled topsoil in Southland, *N.Z. J. Sci.*, 24, 137, 1981.
11. Darling, A. R., The Effects of Native Hay Mulch on Stabilization and Revegetation of Strip Mined Lands in Southeastern Montana, M.S. thesis, Montana State University, Bozeman, 1983, 74.
12. Reclamation Research Unit, Rapid Soil Stabilization, A Roadside Development Summary Report, U.S. Department of Transportation and Montana Department of Highways, Bozeman, MT, 1981, 6.
13. Sloan, J. P. and Ryker, R. A., Large Scalps Improve Survival and Growth of Planted Conifers in Central Idaho, Interm. Res. Sta. Res. Pap. INT-366, U.S. Department of Agriculture, Ogden, UT, 1986, 7.
14. Coenenberg, J., Methods for establishment of diverse native plant communities at the Rosebud Mine, in *Proc. Symp. Surf. Min. Reclam. Northern Great Plains*, Agric. Exp. Sta. Res. Rep. No. 194, Montana State University, Bozeman, 1982, B-5–1, 1982.
15. Young, S. A. and Rennick, R. B., *Use of Grazing Cattle During Post Seeding Management of Revegetated Minesoils*, Reclam. Res. Unit, Montana State University, Bozeman, 1983.
16. Mueggler, W. F., Cattle distribution on steep slopes, *J. Range Manage.*, 18, 255, 1965.
17. Goering, J. D. and Dollhopf, D. J., *Evaluation of a Mine Pit Impoundment in the Semiarid Northern Great Plains, Big Sky, Montana*, Agric. Exp. Sta. Bull. 148, Montana State University, Bozeman, 1982, 30.
18. Skousen, J., Wetlands are more than two cattails, *Green Lands*, Spring, 45, 1989.

6

Applied Revegetation Techniques for Common Disturbances

Basic revegetation techniques such as seedbed preparation, fertilization, and seeding are practiced at most disturbances, but different types of perturbations require modification of these final reclamation activities. For example, revegetation at a coal mine and at a drill pad includes these three basic steps, but the size, location, and soil impacts at the two sites requires that other techniques be attempted to economically and expeditiously revegetate them. In this chapter, several common but hypothetical revegetation examples will be reviewed and discussed in detail to illustrate this principle.

WESTERN COAL MINING

Coal extraction in western North America is usually carried out by stripmining (Figure 48) although there are numerous underground mines and a few open-pit mines in this region. Contour mining is rare in the West but both contour and underground coal mining are common in the eastern coal fields. The surface impact of underground mines is minimal, while the visible impact of open-pit or contour mines are greater than those of any other type of mining. Nevertheless, the problems associated with stripmines receive the major concern of environmental groups because this type of mining is highly visible, is widespread, and has such a long history of serious land abuse. In the United States all surface and underground coal mining falls under the Surface Mining Control and Reclamation Act of 1977 (SMCRA). This law has been amended by Congress and modified by the courts, but it is still an example of what can be accomplished on large-scale disturbances when regulatory agencies and mining companies cooperate to achieve a common goal.

Basically, SMCRA states that prior to any disturbance the vegetation on the site of a proposed mine must be inventoried, a postmine land use defined, and revegetation methods described. Furthermore, the

Figure 48 Western coal stripmine.

topsoil resource must be salvaged and placed over recontoured and nontoxic rootzone materials after mining. Plant seeding must be carried out immediately after recontouring and topsoil application, or during the first appropriate season for such practices in the geographic region of the mine after these activities have been completed. The postdisturbance plant community must be composed of diverse, reproducing, native vegetation comparable to the predisturbance plant community in cover, production, and seasonality.

In the eastern coal fields of North America, most mines are found in forested areas or on agricultural lands. In the western regions of the continent, coal mines are found on the prairie and in forested mountainous areas. Although there are categories for agricultural lands and other postmine land uses, the requirements of SMCRA mean that most surface coal mines in the United States, especially those in the West, will be restored to grasslands or forests after stripmining. Lists of species for seeding on these sites vary from one region to another and even within regions. Texts such as the Soil Conservation Service's *Plant Materials Handbook*[1] and Hardy BBT Limited's *Manual of Plant Species Suitability for Reclamation in Alberta*[2] present descriptions of species commonly used on disturbances in western North America. The author also has two companion texts. One, *Perennial Grasses for Revegetation of Disturbed Lands in the Northern Great Plains and the Intermountain Region*[3] and the other, *Forbs, Shrubs, and Trees for Revegetation in the Northern Great Plains and Adjacent Areas*,[4] which may be consulted for descriptions of appropriate species for seeding in these regions. Similar texts are available for most areas in North America.

A typical northern Great Plains coal stripmine revegetation program is presented to illustrate problems and pitfalls in developing an appropriate seed mix and selecting techniques for rehabilitating such a site. The land use goal for this hypothetical mine is livestock grazing and wildlife habitat. The revegetation program takes place on recontoured and topsoiled minesoils. With the exception of the 3:1 slopes of final cuts, stripmine grades are less than 5:1. Federal law (SMCRA) requires a 122-cm (4-ft) nontoxic rootzone on coal mined lands; therefore, both soils and subsoils at this mine must be conducive to plant growth.

Despite the qualifications developed to ensure good revegetation success (land use definition, slope restrictions, topsoil requirement, and nontoxic rootzone materials), several factors inherent to this geographic region affect plant seeding, germination, and growth on coal minesoils. These include:

1. Climatic extremes, including moisture deficits and very high or very low temperatures
2. Soil salinity
3. Trace element levels
4. Legal restrictions

In general, the northern Great Plains are characterized by annual mean rainfall of less than 40 cm (16 in.), with deficits of more than 50% of the annual precipitation relatively common. Approximately one half of the annual precipitation normally occurs as rain or snow during late April, May, and June. No precipitation or only limited showers occur during the months of July and August. During September or October the probability of precipitation increases, and a fall vegetation greenup occurs in most years. If spring rains do not occur or are of limited amounts, drought is certain and seeding failure probable.

Extremes of heat and cold occur in this region. Winter temperatures drop below −34°C (−30°F) in almost every year and occasionally drop below −40°C (−40°F). Snowfall is variable, so the ground may be bare. When the ground is bare the upper 30 cm (1 ft) or more of soil freezes. In July and August temperatures in direct sun exceed 43°C (110°F) and may exceed 46°C (115°F). While neither extreme is enough to prohibit plant growth, a range of more than 80°C is so wide that it places limitations on the number of species that can successfully complete their life cycles in this area.

Agricultural limits for soil salt content are published as 4 to 8 dS/m.[5] Many minesoils on the Great Plains reveal topsoil electrical conductivity (EC) values within or greater than this range. Conductivities from 4 to 7 or 8 dS/m are not inhibitory to plant growth in most semiarid to arid regions unless other factors such as particle size and sodium content compound the salinity problem. Many species that are endemic to xeric regions are adapted to these or even higher salinity ranges.

Elevated soil concentrations of selenium, boron, or low copper:molybdenum ratios present hazards to revegetation programs in some areas of semiarid or arid mining regions. Excesses of selenium and boron are found in a few soils of the Great Plains. Borderline copper deficiencies occur in even more widespread areas. Neither an excess nor a deficiency of selenium hinders plant growth at levels recorded in natural ecosystems, but animals are poisoned by an excess or a deficiency of this element in their diets. On the other hand, boron deficiencies or excesses inhibit the establishment and growth of plants but have no impact on animals at any natural levels reported in plant tissues. Copper deficiencies are most common in alkaline environments, and low soil copper concentrations inhibit both plant and animal performance. Care must be exercised during minesoil construction to prevent the placement of spoils that have phytotoxic elemental concentrations within the rootzone. In addition, surficial soils expressing trace element problems should not be utilized in minesoil development.

Federal and state regulations require that vegetation on coal minesoils be composed predominately of native species. Exotic plants or naturalized agronomic species that are very competitive should not be used in coal minesoil revegetation programs. On the northern Great Plains, species such as crested wheatgrass and smooth brome are so aggressive that they dominate a new minesoil plant community if they are included in seed mixes, even at low percentages. When used in an appropriate setting, however, these species can make a valuable contribution to revegetation of a disturbance.

Planners of revegetation programs must recognize each of the regional parameters that may limit plant germination and establishment. Revegetation programs must also provide for numerous site-specific variations in climate and soils that inhibit or enhance plant performance. All species selected for minesoil seeding must be capable of germination, establishment, growth, and reproduction under the extreme conditions of the region.

As part of the development of the revegetation plan for the hypothetical northern Great Plains minesoil, we must develop a premine inventory. We may assume that the stripmine has a clay-loam surface horizon and a nontoxic, sandy-clay-loam subsoil. Temperature and precipitation regimes are typical of the northern half of the Powder River Basin. The inventory reveals a shrub/grassland plant community dominated by western wheatgrass, needle-and-thread grass, and big sagebrush *(Artemisia tridentata)*. A significant segment of plant cover in the community is made up of prairie junegrass *(Koleria pyramidata)*, threadleaf sedge *(Carex filifolia)*, fringed sagewort *(Artemisia frigida)*, western yarrow *(Achillea millifolium)*, and prickly pear cactus *(Opuntia polyacantha)*. Numerous other minor species are found in different parts of the community. The area to be reseeded varies from a western wheatgrass/Kentucky bluegrass/silver sagebrush *(Artemisia cana)* bottomland, through the dominant community type, to the bunchgrass-dominated soil of the hilltop plant community. This vegetation type (the hilltop community) was composed of bluebunch wheatgrass *(Agropyron spicatum)*, needle-and-thread grass, and skunkbush sumac. As normally occurs in premine inventories, all three vegetation types were grouped into one community: the shrub/grassland community type.

A minimum of three distinct seed mixes should be seeded on this site. Topsoil should vary from very little or none on the hilltop to 60 cm (24 in.) or more in the ephemeral stream bottom. At most coal stripmines the entire area receives the same depth of a homogeneous topsoil. Only one seed mix would be seeded on this slope. Our example, while less desirable, follows the dominant practice in the region today. That is, one seed mix composed of numerous species.

The seed mix for this stripmine is presented in Table 23. It is an example of the type of seed mix referred to in Chapter 5 as a multiple species mix designed to develop numerous plant communities on the minesoil. At mines at which seed mixes are utilized that are similar to the one illustrated in Table 23, the cool-season grasses and fourwing saltbush *Atriplex canescens* would probably be drill seeded. This seeding technique maximizes protection of the seed from desiccation, predation, and elevated temperatures. The warm-season grasses, the forbs, and the other two shrubs would be broadcast seeded and pushed into protected cracks and crevices in the soil by dragging some type of chain, log, or other device behind the seeder. The two seedings may be accomplished in one or two passes. Both seeding techniques (one or two passes) are used successfully on the northern Great Plains.

The plant species selected for this disturbance are widely distributed in this region and are commonly included in revegetation programs. They contribute to the development of several communities on the site; however, as previously noted, greater diversity could be achieved by designing several seed mixes for different slope positions, aspects, and exposures, instead of one multiple species mix for all habitats on the mine.

The cool-season wheatgrasses and green needlegrass are the backbone of revegetation programs on this type of site. If a seed source for needle-and-thread grass were available, this species would replace

TABLE 23 Seed Mix for a Hypothetical Northern Great Plains Coal Minesoil

Plant Lifeform	Common Name	Seed Rate[a]		Seed Mix[b] (%)
		No./m²	kg/ha	
Perennial				
Cool season grasses	Western wheatgrass	20	0.8	11.5
	Bluebunch wheatgrass	15	0.6	8.6
	Green needlegrass	20	0.5	7.2
	Slender wheatgrass	15	0.5	7.2
Perennial				
Warm season grasses	Sideoats grama	15	0.4	5.7
	Little bluestem	15	0.3	4.3
	Prairie sandreed grass	15	0.2	2.9
	Indian ricegrass	15	0.4	5.7
Forbs				
	Purple prairieclover	20	0.3	4.3
	Blue flax	15	0.2	2.9
	Cicer milkvetch	15	0.5	7.2
	Western yarrow	15	0.03	0.4
Shrubs				
	Big sagebrush	10	0.02	0.3
	Fringed sagewort	15	0.02	0.3
	Fourwing saltbush	10	1.8	25.8
	Winterfat	10	0.4	5.7
		240	6.97	100.0

[a] PLS — $No./m^2 \div 10.8 = No./ft^2$; $kg/ha \times 0.9 = lb/a$.
[b] By weight.

green needlegrass, or at least be added to the mix as a major component of seed mixes in this region. Green needlegrass assumes the role of a dominant species only on the finer-textured soils of the Northern Plains. Sideoats grama *(Bouteloua curtipendula)*, little bluestem *(Schizachyrium scoparium)*, and Indian ricegrass *(Oryzopsis hymenoides)* are often added to coal minesoil seed mixes in this region to provide warm-season diversity to grassland plant communities. Blue grama *(B. gracilis)* is occasionally added to mixes, but it performs better in the central and southern Plains than on the northern Great Plains.

The niche of the warm-season grasses on the northern Great Plains and even in the midwestern coal fields is usually the more arid site: rocky coarse-textured soils, hillslopes, or heavy clay soils where cool-season grasses compete less successfully. On the mild 5:1 slopes of coal minesoils with uniform layers of topsoil, the warm-season species have little chance of persisting in any great numbers with the more competitive cool-season grasses. The continued presence of significant quantities of warm-season species on revegetated landscapes at any disturbance is questionable unless coversoils are selected for this purpose.

On coal minesoils of the northern Great Plains, purple prairie clover *(Dalea purpurea)* and another legume are usually added to seed mixes as nitrogen-fixing components of rangeland plant communities. They aid in the reestablishment of the nitrogen cycle. Prairie clover performs poorly and its nitrogen-fixing powers must be supplemented with an introduced nitrogen-fixing plant species. Introduced legumes [e.g., alfalfa, sainfoin *(Onobrychis viciaefolia)*, sweetclovers (*Melilotus* spp.), and cicer milkvetch *(Astragalus cicer)*] usually disappear from well-managed minesoil pastures in a few years. The forb, blue flax *(Linum perenne)*, and the halfshrubs, fringed sagewort and western yarrow, may be added to the seed mix simply to lend diversity to the resultant plant community. Shrubs such as winterfat *(Ceratoides lanata)*, fourwing saltbush, and one of the sagebrushes may also be added to the mix. They provide diversity and forage for wildlife as well as livestock. Rubber rabbitbush *(Chrysothamnus nauseosus)*, which performs well when seeded on minesoils, could be used in the seed mix to lend diversity and cover to the new plant community; however, it has poor grazing value and is of limited use to wildlife. It should not be substituted for shrubs with greater potential for grazing or wildlife browsing.

Shrubs may be planted rather than seeded, since transplants are much more likely than seedlings to survive the rigors of the climate and competition from other plants. Many western coal mines are also required to plant trees, even on the Great Plains. These are usually planted as containerized or bareroot stock. If few of the shrubs establish and survive, or if diversity is very low, containerized plants of some of the perennial species should be transplanted into the grass stand. Low diversity is evidenced by the absence of forbs, halfshrubs, and shrubs in the community two or three growing seasons after seeding.

Diverse seed mixes such as the one illustrated in Table 23 are rare on degraded sites. Only on coal minesoils is this level of complexity found in seeding schemes. Despite the seeding of these complex mixes, coal mining companies are having an increasingly difficult task of developing diverse plant communities for bond release. How to encourage diversity in plant communities on disturbed landscapes was addressed in Chapter 5 of this text.

METAL MINE WASTES

The term hardrock as used in this text refers to metal mines. Traditionally these mines are underground and in strongly consolidated materials — thus the term hardrock. Because of the volume of ore processed, safety factors, and labor costs, many of these mines are now open pits (Figure 49). These disturbances are usually in

Figure 49 Open-pit metal mine.

mountains but below the timberline. The problems of alpine or tundra reclamation are unique and are not discussed in this text. Associated with open-pit metal mines are several different types of revegetation: these include the open pit itself, waste rock removed from the pit to gain access to the ore, access and exploratory roads, drill pads, and tailings.

Metal mining is often controlled by state and federal regulations that are "limited in their scope and effectiveness".[6] When a mine is on federal lands (lands administered by the Forest Service or Bureau of Land Management [BLM]), the BLM is responsible for issuance of the mining permit. Mining for phosphate, many nonferrous metals including gold and silver, and other minerals is permitted by this agency. Application for mining permits is based on the 1872 Mining Act, its amendments, subsequent legislation pertaining to air, water, and soil pollution, and to regulations concerning leasing, mining, and multiple use of federal lands. None of this legislation directly addresses reclamation of disturbances associated with mining. As a consequence of the absence of direct legislation pertaining to metal mine reclamation, rehabilitation of surface or underground noncoal-mining disturbances on federal or state lands is usually poorly planned, inadequately implemented, and of limited success. The return of federal forest or rangeland to plant production comparable to premine vegetation production potential after disturbance by hardrock mining is rarely seen.

Noncoal permit applications often do not contain a clear definition of the postmine land use or of the methods proposed to achieve that end. Reclamation plans, if required by state or BLM regulations, often consist of a few pages of text discussing the present ecological setting of the mine and the proposed future of the impacted area. The federal or state permit application is rare that contains a full description of the ecosystem to be impacted and how and when the impact will be mitigated. As long as regulatory agencies do not require site-specific and detailed reclamation plans in permit applications, effective reclamation of the disturbance will not be carried out. For example, it is only when mining companies define the length, aspect, gradient, and materials of final waste rock dumps or hillslopes that geomorphologists can ascertain the stability of the slopes and botanists the potential for revegetation. Such data are found in few hardrock mine permits and, when incorporated into the permit, they do not contain the binding authority of coal regulations.

Open Pit

In open-cut mines complete backfilling of the pit usually is not economically feasible. Because the pit remaining after mineral extraction may be several hundred feet deep, the cost of moving waste rock back to the depression is prohibitive. Surface and ground water usually flow into the depression after mining is terminated and an impoundment develops. These water bodies may be assets, but some degree of landscaping is necessary before they can reach their full potential. Pit slopes should be reduced to the limits required for safe access for humans, livestock, wildlife, and for the operation of revegetation equipment on the contour. Pit slopes should not be permitted at gradients that jeopardize the success of postmine reclamation.

Waste rock removal programs should be coordinated with mineral extraction to the maximum extent possible when large open-pit mines are constructed. Partial backfilling of the pit is possible in the majority of such mines. Material handling costs can be curtailed and the size of the disturbance reduced if ore extraction and waste rock movement are planned and coordinated.

Topsoil is rarely available for redistribution on pit slopes. Rehabilitation must often be carried out in rock rubble. Revegetation of open-pit slopes and associated disturbances is identical to the revegetation of waste rock. The problems, activities, and seed mixes described in the next section pertain to the open-pit disturbance as well as waste rock.

Waste Rock

Hardrock mines in western North America are usually in mountainous terrain. Although some mines are above the timberline, most of these disturbances are below the alpine ecosystem. This discussion of waste rock revegetation is restricted to disturbances at high elevations but below the timberline in the subalpine. Undisturbed coarse textured surface soil materials of these regions often reveal few fines but have high organic matter content. Cation exchange capacities (CECs) are adequate because of the high organic content of the surface layer. Beneath this surface, however, is a nutrient-poor layer composed almost exclusively of lightly weathered to unweathered parent rock. Revegetation of this waste rock after removal of the surface layer poses a number of unique problems. Chief among these potential problems are

1. The steep slope of the material
2. Coarse nature of the waste rock and resultant low CEC, low water holding capacity, and poor nutrient cycling potential
3. Absence of organic matter
4. Climatic extremes including very high and extremely low temperatures, wind, and drought
5. Fertilizer, seed, and mulch incorporation are very difficult
6. If a cover soil is available for deposit over the waste rock erosion may be severe
7. Acid generation from reduced sulfur-containing materials

The construction of an open-pit mine and to a lesser extent underground mines in mountainous terrain requires the removal of large quantities of rock. This coarse-textured spoil material is dumped in head-of-hollow or valley fills. The angle of repose of these materials (gradients are steeper than 2:1) is not adequate to ensure long-term revegetation success, even if topsoil is available for distribution over the face of the dump. The reclamation specialists must reduce the slope of the dumps. This may be accomplished by constructing the waste pile in lifts and leaving terraces between successive lifts, or by recontouring the face of the waste pile after termination of the dumping program.

Waste rock is very coarse (Figure 50) — commonly with 40% of the material greater than 5 cm (2 in.) in diameter. Even in the fraction smaller than 5 cm the particles tend to be in the larger size categories. For example, in a study at a gold mine in north-central Montana, it was reported that on a weight basis 83% of the particles less than 5 cm were greater than 2 mm in diameter, and on a volume basis only 15% of the material was less than 2 mm.[7] Such material has low CECs and very poor nutrient status. Mineral cycling cannot take place

Figure 50 Waste rock at a western gold mine.

in these materials until a pool of organic matter is established and the bacteria, fungi, and arthropods necessary for the decomposition of organic materials establish populations in the surface rock/mulch layer.

Infiltration rates are rapid on waste rock slopes but water-holding capacity is low; therefore, the sites have periodic droughts. While often very steep, they are not as erosive as similar slopes in other materials because they lack fines and have high infiltration rates. These properties of waste rock change if the slopes are adequately covered with surface soils or composed of rapidly weathering waste rock.

Mountains are generally subject to moisture, wind, and temperature extremes. During spring precipitation and snow melt, western North American mountains are richly supplied with water, but as soon as this spring flush of moisture has passed a prolonged drought ensues. Temperature extremes are colder than in the plains but usually not so hot. Higher elevation sites are also subjected to frost almost every month of the year and, while the day length is long, the growing season is very short. Strong winds are characteristic of higher mountainous slopes. These winds and the intense solar radiation that usually accompanies them scours any soft vegetation that is exposed and dehydrates plant tissues. These extreme climatic conditions make waste rock revegetation difficult, even under less demanding conditions of slope and texture. Combined with the physical restrictions of coarse-textured materials and low water-holding capacity, revegetation of waste rock slopes becomes very difficult.

To establish fertilizer amendment rates, the macronutrient status of topsoil and waste rock must be determined. Fertilization is indispensable when topsoil is not available. Hardrock mine wastes are deficient in nitrogen, phosphorus, and possibly potassium. Amendment of these materials with macronutrients is necessary. It is imperative that nutrient additions be undertaken as soon as possible on waste rock to support plant growth and reduce erosion. At the vast majority of mountain disturbances, one heavy application of fertilizer is not adequate. Fertilizer applications over several years are necessary to enhance plant growth, to build up the organic contents of the surface horizon, and to initiate the cycling of nutrients in these materials.

The value of surface coverings more amendable than waste rock to plant germination, establishment, and growth has become clearer during the past few years. In the absence of such a material, a combination of sewage sludge and fly ash has been successfully used as a topsoil substitute in Pennsylvania.[8] Compost, in combination with other amendments, has also been successfully used on waste rock in Montana.[9] Soil construction with these materials is discussed in Chapter 3.

Plant selection is more critical for higher elevation waste rock revegetation programs than it is for coal minesoils. Species that thrive in nutrient-poor mineral soils must be utilized even when topsoil is available because nutrient deficient waste rock makes up the subsoil and a major portion of the rootzone materials. Some sort of broadcast seeding is necessary unless slopes are substantially reduced and topsoiled. Seed and fertilizer should be incorporated into the seedbed. This is difficult because of the slope and size of the materials that constitute the seedbed. Because other methods are so unreliable, hydroseeding and hydromulching are often practiced on waste rock dumps. These two techniques are appropriate on this material but both are limited by the area the equipment can cover from one position. If large expanses of waste rock are to be seeded and mulched by this equipment, special access must be provided to permit movement of the hydroseeder or hydromulcher over the disturbance.

If fines are available for distribution over the coarse fragments, or if topsoil is salvaged for respreading over the waste, rapid revegetation is possible and may be relatively easy, but other factors can seriously complicate the program. Aridity is one of these problems. Runoff is accelerated by the steep face of the slope and the water-holding capacity of mountain soils is low. Drought is common on such hillslopes.

Erosion may also be extremely severe on topsoiled slopes. Many bare slopes with gradients steeper than 3:1 are unstable when saturated. Slumping, sheet erosion, and gully formation must be a concern if any significant amount of topsoil (several centimeters) is placed over waste rock on steep slopes (>33%).

The majority of hardrock mines are in the western mountains, and the dominant vegetation on such sites prior to disturbance is coniferous forest. Pines, firs, and spruce are harder to establish from seed than grasses and forbs. On the steep slopes of waste rock dumps, it is easier to establish grasses and forbs for erosion control and aesthetics than it is to return the landscape to its previous vegetation cover. Mountain grasslands are common, and species composition for many of these vegetation types has been documented.[10] This information and premine inventories identify grass species for seeding on these slopes. Organic matter accumulation following grass sward establishment increases the cation exchange and water-holding capacities of the soil and permits the invasion of adjacent woody plants at some later date.

Two hypothetical seed mixes for topsoiled waste rock dumps are presented in Table 24. No plant diversity standards exist for hardrock sites; therefore, these seed mixes are not as complex as those constructed for coal minesoils (Table 23). Native species predominate in the first mix. They are adapted to the climate of such sites in the western mountains and they meet the requirements of the most commonly designated land use (livestock grazing and wildlife habitat). The second seed mix is more typical of those used for many years by the Forest Service on clear cuts and burned areas from the East to the West coasts. These introduced grasses germinate, establish well, and produce abundant forage but they are replacing many native species on mountain grasslands throughout North America. How these changes in plant communities affect big game and nongame wildlife, long-term livestock carrying capacity, and timber regeneration is unknown. Fortunately, these seed mixes are being discarded for native species seed mixes in most forest districts.

The advantages of coversoils have been demonstrated in the coal fields,[11,12] but little research has been carried out on the use of coversoil or topsoil at higher elevation waste rock dumps.[13,14] It must not be assumed, however, that coversoils are indispensable to successful

TABLE 24 Two Seed Mixes for a Hypothetical Topsoiled Waste Rock Dump in the Semiarid West

Plant Lifeform	Common Name	Seed Rate[a] No./m²	Seed Rate[a] kg/ha	Seed Mix[b] (%)
	Native Species Mix			
Perennial				
Cool season grasses	Bluebunch wheatgrass	60	2.3	18.9
	Hard sheep fescue	60	0.5	4.1
	Mountain brome	40	2.2	18.1
Forbs				
	Yellow sweetclover	10	0.2	1.6
	Common lupine	20	6.9	56.6
	Western yarrow	30	0.04	0.3
Shrubs				
	Fringed sagewort	20	0.02	0.2
	Big sagebrush	10	0.02	0.2
		240	12.18	100.0
	Introduced Species Mix			
Perennial				
Cool season grasses	Timothy	300	1.2	21.1
	Orchard grass	300	2.4	42.1
Forbs	Red clover	120	2.1	36.8
		720	5.7	100.0

[a] PLS — No./m² ÷ 10.8 = No./ft²; kg/ha × 0.9 = lb/a.
[b] By weight.

waste rock rehabilitation. Many high elevation grasses and forbs are adapted to mineral soils and perform better on these materials than on topsoiled minesoils. Several varieties of grasses and forbs have been successfully seeded on bare waste rock.[15]

More recently, the seeding of native grasses, forbs, and shrubs on all disturbances has been recommended by the National Park Service.[16] The program in place in Yellowstone National Park is utilizing native seed stock collected within the immediate neighborhood of the disturbance to be restored. The goal of this program is to create wildlife habitat and aesthetically pleasing landscapes, control erosion, and maintain the genetic integrity of native species in the vicinity of the disturbance, while rapidly returning lands to plant production. These seedings have been very expensive but also very successful. Programs utilizing entirely native seed mixes in the coal fields and in Yellowstone National Park call into question the indiscriminate use of introduced species on so many other disturbances.

Two hypothetical seed mixes composed primarily of native species adapted to bare mineral soils are shown in Table 25. The plant species noted in this table are found on high-elevation, coarse-textured, low-fertility soils. They are relatively productive and provide good wildlife

TABLE 25 Two Seed Mixes for a Hypothetical Waste Rock Dump Without Topsoil in Semiarid Western North America

Plant Lifeform	Common Name	Seed Rate[a] No./m²	Seed Rate[a] kg/ha	Seed Mix[b] (%)
	North or East Slope			
Perennial				
Cool season grasses	Sheep fescue	60	0.4	4.3
	Bluebunch wheatgrass	60	2.2	23.9
	Prairie junegrass	30	0.1	1.1
	Columbia needlegrass	30	0.8	8.7
	Mountain brome	20	1.1	12.0
Forbs				
	Western yarrow	20	0.1	1.1
	Common lupine	10	3.5	38.0
Shrubs				
	Snowbrush ceanothus	20	1.0	10.9
		250	9.2	100.0
	South or West Slope			
Perennial				
Cool season grasses	Hardsheep fescue	80	0.6	5.6
	Bluebunch wheatgrass	80	2.9	27.1
	Sandburg bluegrass	30	0.1	0.9
	Indian ricegrass	30	0.9	8.4
	Slender wheatgrass	30	1.0	9.4
Forbs				
	American vetch	30	4.5	42.1
	Western yarrow	30	0.1	0.9
	Blue flax	30	0.4	3.7
Shrubs				
	Fringed sagewort	40	0.1	0.9
	Big sagebrush	40	0.1	0.9
		420	10.7	99.9

[a] PLS — No./m² ÷ 10.6 = No./ft²; kg/ha × 0.9 = lb/a.
[b] By weight.

forage and livestock grazing even on waste rock without topsoil. The first seed mix was created for a 1500-m (5000-ft) or more, north-facing, waste rock slope in eastern Washington, Idaho, or Montana and the other for a south-facing slope in this same region. The Idaho fescue/bluebunch wheatgrass community is one of the more common habitat types of the mountains west of the northern Great Plains and these mixes resemble the composition of this community. Sheep fescue *(Festuca ovina)* and hard sheep fescue *(F. ovina* var. *duriuscula)* are substituted for Idaho fescue *(F. idahoensis)* because they germinate and establish better.

The north-facing slope is more mesic and includes Columbia needlegrass *(Stipa columbiana)*, mountain brome *(Bromus marginatus)*, and lupine (*Lupinus* spp.), all of which require a slightly more mesic environment than is available on south-facing slopes. The yarrow and ceanothus *(Ceanothus velutinus)* provide diversity and wildlife habitat to the site. The lupine is a native nitrogen fixer. Nevertheless, sheep fescue and bluebunch wheatgrass should provide the backbone of the plant community on this site.

The second seed mix includes slender wheatgrass *(Agropyron trachycaulum)* as a nurse crop and a higher seed rate. Seed rates are augmented on the south-facing slope because seed and plant loss on harsh exposures is high. Bluebunch wheatgrass and hard sheep fescue will share dominance on this slope, but blue flax and sagebrush will be common. If sagebrush does not establish well following seeding, tubelings of the Hobblecreek cultivar of Mountain Big Sagebrush *(Artemisia tridentata* var. *vaseyana)* should be planted in the spring of the second or third growing season. Yellow sweetclover (*melilotus officinalis*) is added to the seed mix only as a nitrogen-fixing legume in the plant community.

Seed mixes used on waste rock dumps are simple aggregations of propagules of plant species selected to meet the average soil and environmental factors at the disturbance. One seed mix is often used on numerous small drill pads or several large waste rock dumps. The disturbances may be located at elevations from a few hundred meters (yards) above sea level to the alpine, and on aspects from north to south. More efficient plant establishment and growth can be obtained when seed mixes are designed for each disturbance. In mountainous terrain environmental conditions change very rapidly and seed mixes should be modified to reflect these changes.

Access Roads

A maze of roads has been constructed in the western mountains for exploration, for access to mineral deposits and drill pads, and even for timber cuts. These access roads are usually not as steep as the slopes of waste rock dumps but they produce the same handicaps to plant growth. In addition, rights-of-way have unique characteristics that further reduce the probability of successful plant establishment. The site parameters that inhibit plant germination, establishment, and growth on access roads in the western mountains are

1. The coarse nature of the materials and resultant low CEC, low water holding capacity, and poor nutrient cycling potential
2. The absence of organic matter
3. Climatic extremes including temperature fluctuations, winds, and drought
4. Difficult fertilizer, seed, and mulch incorporation
5. Severe erosion potential
6. Compaction

Most of the discussion of waste rock revegetation is pertinent to access road rehabilitation. Development of an organic surface layer and nutrient cycles is imperative for long-term vegetation persistence. Roads are avenues for the incursion of noxious weeds carried on vehicles or livestock into previously undisturbed native plant communities. Fast establishment of permanent vegetation on the roadbed is the best deterrent to the invasion of noxious weeds into virgin territory.

The climatic extremes noted on waste rock are active on access roads. Violent temperature fluctuation, wet spring and droughty summer weather patterns, and strong winds hamper plant establishment and growth on these sites.

Compaction can be alleviated by ripping. This increases infiltration and permits better root penetration.[17] In addition, ripping provides a limited avenue of organic matter and fertilizer incorporation. These two amendments are indispensable for rehabilitation of access roads when surface soil is not available and they enhance plant establishment and growth when topsoil is available.

If surface fines were pushed off the roadbed during construction and remain as berms on the shoulders, they should be pulled back onto the bed. If this surface material is not available, any soil resource can facilitate revegetation. In the absence of a surface soil resource, which is more common when revegetating mountain access roads, sewage sludge or compost may be substituted for a seedbed.

Surface construction must include some provision for erosion control. Long roads on slight grades (5 to 20%) become conduits for water movement and soil erosion after ripping, amendment, and addition of a cover soil. Recontouring, and the use of diversion structures to limit the buildup of water volumes, must be part of the rehabilitation program.

Seeding of topsoiled roadbeds may be carried out by drill or broadcast seeding. Nontopsoiled but ripped sites may be broadcast seeded and followed by cultipackers, chains, or some type of device to press seeds firmly into the seedbed or push them into cracks and crevices where they are protected until germination. Seed mixes used for access roads should vary locally by elevation, exposure, and substrate. The second seed mix in Table 25 is appropriate for a high elevation access road with a south or west exposure, although mixes based upon local vegetation would be more appropriate. The second mix in Table 24 is more representative of historical seedings in the western mountains. The first seed mix in Table 24 and Table 25 represent two mixes that could be used as readily on access roads as on waste rock dumps.

Before revegetation of access roads can be considered successful, road closure is a necessary requirement but one that is difficult to enforce. Ditching and steel gates are poor deterrents to the off-road vehicle enthusiast. The transplanting of vigorous trees into the roadbed helps create the illusion of permanent closure and nonuse. Planting of tubelings or bareroot stock of shrubs is a revegetation technique that strengthens this illusion.

Tailings

Mill tailings or smelter wastes are the materials remaining after processing has removed the mineral from an ore. Because of their chemical and/or physical properties, they inhibit plant and microorganism growth, provide a serious ground and surface water pollution source, and are aesthetically unpleasant. Tailings disposal is carried out by burying the materials back in the mine; dumping them into the ocean, a stream, or lake; or by placing them in a receiving pond. Many streams are now toxic waste sites because for many years they were the dumping ground for smelter wastes. In New Guinea and Chile, rivers are still the repository for tailings despite the environmental damage incurred by this practice. Lake Superior was long the recipient of tailings from taconite processing plants in Minnesota. Tailings disposal is strictly controlled in the United States today. Uniquely constructed and sealed ponds are required for the permanent entombment of these materials. The question of the need for these storage structures is no longer relevant. Discussion now revolves only around what is adequately isolated.

Tailings storage areas are not desirable additions to the landscape because of their toxic contents. Therefore, they are placed out of sight, usually in inhospitable climatic regions. Large expanses of mill tailings are found in South Africa, much of Australia, and western North America. Major North American tailings disposal areas are found from central Canada to Tucson, Arizona. With the growth of the population of western North America, what was once a minor out-of-sight annoyance is now a serious pollution source.

Tailings exhibit numerous characteristics that limit plant germination, establishment, and growth. Chief among these are

1. High salt contents
2. Either acidity or alkalinity and accompanying elevated toxic metal levels and macronutrient deficiencies
3. Particle size stratification
4. Lack of organic matter

Because of their location, tailings ponds are also subject to climatic extremes including high winds, elevated temperatures, and drought.

Tailings are usually sluiced to the holding pond (Figure 51). Plant-inhibiting salt levels in the tailings and in the liquid carrier are often high. They are a residue of extracting agents, salts in the ore itself, or salts formed as a result of the reaction of lime and acid in pyritic tailings. The salinity of these materials is exacerbated by evaporative losses and a concentration of salt in surface layers as they dry. Salt contents in tailings may range from 0.1 to 22 dS/m.[18,19] For all mill wastes — acid, neutral, or alkaline — salinity adds another level of difficulty for revegetation.

Wastes from the processing of ores may have a final pH that ranges from very acid (pH = 1.8) to very alkaline (pH = 9.4),[20,21] either because of the nature of the ore or the addition of various agents during the extraction of the mineral. The common pyritic ores may be alkaline upon discharge to the receiving pond but they gradually acidify, whereas wastes from oxidized ores may remain in the holding pond at a relatively stable pH. The establishment of vegetation on neutral tailings is less difficult than revegetation on alkaline or acid materials. Although alkaline tailings do not express all of the phytotoxic chemical properties inherent in acid wastes, both acid and alkaline tailings are very difficult to revegetate. While the following discussion pertains directly to acid tailings, it is also appropriate to acid-generating mine wastes other than tailings.

Figure 51 Tailings being sluiced to a holding pond.

Reduced sulfur-containing materials (e.g., pyrite and marcasite) in tailings or waste rock are oxidized when exposed to air and water. Acid generated by the oxidation reduces pH and increases the solubility of metals. Concentrations of aluminum, copper, manganese, and zinc often reach levels that are toxic to plants and microbes. In very acid wastes (e.g., pH <4) it is often questionable if the toxicity is a result of the excess of hydrogen ions, a deficiency of bases on the exchange, or metal toxicities. At most tailings or acid waste sites, all three phenomena are operating simultaneously. Other metals (e.g., cadmium, lead, and nickel) rarely reach phytotoxic concentrations in tailings, but analyses often indicate that levels in the materials pose threats to animals and man. Problems of food chain accumulation of toxic metals must be evaluated during the revegetation phase of tailings rehabilitation.

Aluminum, copper, zinc, and other metals may pose threats to man and animals via several pathways. The avenues to animals include plant uptake and animal consumption of contaminated plant tissues. Dust or dirt particles adhering to plant leaf surfaces exacerbates the danger posed to herbivores from consumption of vegetation containing elevated trace metal levels. For some metals (e.g., arsenic, cadmium, and lead) this is a more severe threat to animals than plant uptake. Direct intake via consumption of dirt while grazing or browsing poses a threat to livestock and wildlife. Direct human intake can be controlled by institutional constraints such as fencing, but wildlife have little regard for such structures.

Dust blowing from these sites and its inhalation by animals, including man, poses a very subtle and continuing threat to man or animals living adjacent to tailings ponds. Two temporary remedial activities are practiced. They are capping with a coarse nonerosive material, and flooding. Waste rock is a common capping material. When the rock is limestone it may react with the acid tailings to form a hard surface crust that resists suspension by wind blowing across the pond. Water is also held on tailings ponds to inhibit the movement of tailings fines by wind. Control of water levels and fertilization of tailings permits the growth of acid-tolerant grasses and other plants on irrigated wastes. Grasses such as tufted hairgrass and redtop *(Agrostis alba)* readily colonize acid sites in temperate regions of the world.

In general, toxic metal concentrations and acid levels in tailings must be reduced before microorganisms can develop and most plants can germinate, establish, and grow in the materials. Neutralization of the acidity of mine tailings reduces the availability of metals. Typically, this is accomplished with limestone ($CaCO_3$), but other calcium compounds may perform this function better than calcium carbonate.

These other calcium sources include compounds such as calcium hydroxide ($Ca[OH]_2$) or calcium oxide (CaO). Industrial wastes such as power plant fly ash or cement plant kiln dust also contain calcium. The neutralization potential of these wastes is low and the quantities needed for treatment so large that transportation charges make their use prohibitively expensive. Where a source of these materials is readily available, however, they may be substituted for lime.

The main problem associated with the neutralization of acid tailings with any calcium-containing material is the determination of the quantity of amendment needed. Application of excess calcium is expensive and contributes to the salinity of the wastes. Application of too little amendment necessitates another treatment or treatments in the future. The correct amount of amendment is equal to the amount of base necessary to neutralize the sum of the hydrogen ions in the soil solution, hydrogen and aluminum ions on exchange sites, and residual acid production capacity in the waste. Acid-base accounts are the most common method for determining the amount of base necessary to neutralize an acid-generating material. This analysis is based on a summation of the various forms of acid-generating compounds and the determination of the neutralization capacity of the waste. The acid-generating potential in excess of the neutralizing capacity is the amount of lime needed to neutralize the waste material. This latter term is difficult to calculate. For a more complete discussion of this topic the reader may consult Caruccio and associates,[22] Sobek et al.,[23] or other worthwhile studies of acid-generating mine wastes.

Tailings are almost always deficient in available nitrogen, phosphorus, and potassium.[24] Acid-generating tailings are also low in calcium. Amendment with lime (calcium compounds) raises the pH and increases the calcium concentration. This reduces trace element availability, including the availability of boron, but it has little influence on the macronutrient content of the wastes. An initial fertilization with nitrogen, phosphorus, and potassium is necessary to stimulate microorganism activity and to start nutrient cycling.[25] Subsequent fertilization may be required to support plant growth until nutrient cycling in the tailings can supply adequate macronutrients to the higher plants.

Boron fertilization, if necessary, is required only once. As with other degraded soils, fertilizer additions should be based on analyses of the seedbed and near-surface soil. Unfortunately, there are no definitive methods for the analysis of nutrients in these types of wastes. Standard agricultural techniques of nutrient analysis do not supply satisfactory data in these materials.

In addition to harsh chemical properties, tailings have physical characteristics that make them unsightly and inhibitory to plant life.

The materials are usually carried to the site suspended in water (Figure 51). When tailings are dumped into the holding pond, particles settle out by gravity, with the larger particles dropping nearer to the effluent opening and smaller particles being deposited at progressively greater distances from the opening. This differential settling pattern produces layers of permeable horizons and impermeable materials alternating throughout the tailings storage pond as the outlet for the waste material is moved around the basin. The smaller particles are silt to clay size and form slimes that tenaciously hold water. Tailings ponds, therefore, are difficult to dewater after decommissioning and have poor infiltration rates. Some tailings hold water so tenaciously, even after drainage and dewatering, that they form a jellylike structure which turns to a liquid when agitated. Such materials are referred to as thixotropic. Their surfaces are unsafe for mechanized equipment for months to years after closure.

Tailings are devoid of organic matter and microorganisms. Tailings result from the grinding of rock and deposit of the fine material as a slurry in some type of holding area. The ores lack organic matter in their original state and none is added in the extracting process. Nutrient cycles cannot be established until organic material has accumulated and microorganisms are present and active in the wastes. This deficiency hinders long-term revegetation success. Both compost and sewage sludge are valuable organic additions to this type of waste. They add the organic matter necessary to improve rootzone properties and the microorganisms necessary to establish healthy nutrient cycles.

Tailings surfaces are the color of the ground material plus any supplements added to assist in the removal of the mineral from the ore. After they are sluiced to the holding pond they are exposed to direct solar radiation. Dark tailings absorb the long wavelengths of incoming radiation and become very hot. In temperate regions, summer temperatures of tailings surfaces exceed 50°C (122°F), the lethal soil temperature for trees.[26,27] At these temperatures any plants that grow on the surfaces are killed. Light-colored tailings (Figure 52) reflect most incoming radiation. The reflected radiation is so intense that it may burn the underside of leaves of plants growing on the wastes. Shading, therefore, often becomes a requirement for the stabilization of tailings with vegetation.

The surfaces of tailings ponds are often subjected to strong winds because they are flat, located in exposed locations, and of large size, sometimes over 1000 ha (2500 acres). Tailings particles are so easily picked up by air currents that plants growing on or near tailings impoundments are sand blasted and defoliated. The stressed plants are killed by the wind and its sediment load. Seeds are also blown

Figure 52 Dry, light-colored tailings.

away and seedbeds buried by material carried on strong local winds blowing over tailings ponds. Wind protection can be incorporated into the shading program during revegetation. Mulches, erosion control blankets, and snow fences, among other techniques, can provide some degree of wind protection on tailings ponds.

Revegetation programs for acid-generating tailings must incorporate multiple actions simultaneously. That is, several major chemical characteristics of the rootzone materials as well as physical and climatic properties of the site must be addressed at the same time. Liming must be implemented and time allowed for the amendment to neutralize adverse soil chemical properties. However, before or at seeding, nutrient supplies in the rootzone must be amended, organic material must be added, shading must be provided, and wind controlled if vegetation is to persist. Organic amendments (e.g., manure, wood chips, and compost) may be incorporated with the lime or at seeding. This helps alleviate soil properties such as compaction, poor infiltration, and lack of organic matter if incorporated with lime. Despite these amendment programs, vegetation selected for seeding or planting on acid tailings ponds should be tolerant of acid and high metal levels. Acid-tolerant vegetation reduces the need for lime and is a precaution against poor mixing of amendments with acid wastes. All vegetation seeded on tailings must also tolerate nutrient-poor soils, elevated salinity, and the climatic regime of the area. Even with these precautions acid tailings revegetation programs often fail.

Failure is generally caused by incorrect amendments or amendment rates, salinity, low macronutrient concentrations, poor plant species selection, or the absence of provisions for the control of adverse physical conditions on the wastes (e.g., wind and temperature extremes).

The basic problem of acid tailings rehabilitation is construction of a soil profile over the waste that is thick enough to support long-term plant growth and prevent acidification or salinization of the soil cover. Data from investigations of soil depth over acid-generating overburden showed increased plant production with increased depth of soil.[28] The deepest soil in this study was 152 cm (60 in.). At this depth, plant production was approximately 85% of that measured at the site prior to disturbance. It may be inferred from these data that some soil depth greater than 152 cm is necessary to maintain postdisturbance plant production levels comparable to undisturbed sites around the tailings pond. Studies of salt migration from coal spoils into coversoils indicate that approximately 70 cm (28 in.) of good quality topsoil are necessary to prevent salt migration into surface soil layers.[29,30] It is apparent that a nontoxic rootzone over 70-cm thick (and possibly over 150 cm) is required to support a self-regenerating and diverse plant community on acid mine tailings.

Revegetation of acid-generating tailings ponds usually includes dewatering and covering with a soil mantle. At many tailings ponds the waste is sealed with a compacted clay cap (20 cm [8 in.] deep) to prevent water and oxygen from infiltrating the tailings. This prevents acid formation, but 20 cm of compacted clay also prevents plant roots from penetrating the cap. The application of 50 cm (1.5 ft) or more of overburden, subsoil, or rock over the clay cap, and a cover soil of up to 30 cm (1 ft) or more over the overburden completes the traditional acid tailings reclamation process.[31] Construction of the cap and a rootzone over the cap is a major expense. Deeper soil covers require correspondingly greater capital investment.

When tailings are covered with an adequate rootzone of nontoxic soil, seeded vegetation must be adapted only to the new rooting materials, climate, and elevation of the site. Although seed mixes vary by location and soil properties, seedings for such materials could be represented by the mix found in Table 23 and both mixes in Tables 24 and 25, varying with elevation, soil, and location of the tailings. Permanent stabilization of acid-generating tailings covered with a cap and nontoxic rootzone materials may be accomplished with normal agricultural practices of seedbed preparation, fertilization, seeding, and management. Temporary irrigation is necessary only in the driest regions.

Rehabilitation of acid-producing materials is usually much less expensive than the example noted above. Tailings or any acid-generating wastes are commonly dewatered, a cap (of clay or limestone) placed over the waste, and a thin layer (40 to 50 cm [16 to 20 in.]) of subsoil and/or coversoil placed on the cap. At other sites the surface of the wastes may be amended with some type of calcium compound to neutralize acid production then a cap and soil placed over the amended tailings. These soil layers are commonly 40- to 50-cm (16- to 20-in.) thick. Revegetation of these types of sites is temporarily successful, and one or two species of acid-tolerant grasses may persist indefinitely,[32] but the continued existence of multiple plant species in the cover is doubtful in arid to semiarid regions. Development of a complete, self-perpetuating, and diverse plant community in thin soil layers will not occur. Plant roots do not penetrate limestone covers or acid materials and a 40- to 50-cm rootzone is not thick enough to support long-term growth of multiple plant species in dry regions. Even a 50-cm rootzone over an impermeable clay cap is inadequate for all but a small number of shallow-rooted species.

The total soil layer over any acid-generating waste or tailings must be thick enough to prevent salt or acid migration into the surficial soil layer and maintain a permanent nonacid and nonsaline rootzone. Seeding and fertilization are carried out in the soil covering. The plants that grow on the site must produce sufficient below-ground biomass to harvest all precipitation infiltrating into the cover soil to prevent continued acid formation in deeper wastes. Long-term (many years) success or failure of tailings revegetation programs is usually determined by the acid-generating capacity of the waste, amount of amendment added to the waste, total thickness of the rootzone, and plant selection.

The use of diffusion barriers is being studied on mineral processing waste ponds in the West.[33] These layers of coarse-textured materials are placed between the tailings and the coversoil. Under the influence of gravity, the barriers freely pass water to lower horizons. They do not impede the growth of plant roots, but they do hinder the diffusion of acid or salts from lower layers into surface coverings. When this type of rehabilitation program is instituted on wastes, the total rootzone includes the coversoil, the diffusion barrier, and any part of the tailings that may not be phytotoxic. This allows the revegetation specialist to rehabilitate these harsh sites with thinner soil coverings. The combination of amended tailings, diffusion barriers, and thinner coversoils may help reduce the very expensive soil handling costs of tailings revegetation.

The successful rehabilitation of acid tailings without soil coverings has not been reported unless the site was leached before seeding.[34] Saline/alkaline tailings reveal similar rehabilitation problems.[35] In semiarid and arid regions sufficient soil material to cover acid tailings to an adequate depth rarely occurs. At such sites construction of a plant-compatable rootzone that is deep enough to support a diverse plant community and that will not acidify is at the heart of the reclamation plan. However, determination of the amount of amendment (lime) and mixing that lime to a depth that provides a rootzone adequate for permanent vegetation growth is difficult.

The reclamation of pyritic mine waste using sewage sludge, agricultural lime, and fertilizer without a coversoil was demonstrated in the eastern U.S.[36] The trial successfully produced a grass cover on the study site, but annual maintenance (spot application of lime and fertilizer) is required to maintain the quality of the vegetation cover. At this site, the amount of lime amendment was not sufficient to neutralize all of the acid-generating capacity of the waste and the layer of amended material was not thick enough to prevent migration of acid into surface layers. Nevertheless, the amount of lime necessary for neutralization of all of the acid-generating capacity in the rootzone can be calculated[37] and the amendment can be mixed with the tailings to a depth of almost 1 m.[38] Within a few years, the depth of incorporation will be increased as new heavy-duty equipment is developed. The very expensive movement and spreading of coversoils over tailings can be avoided using these reclamation techniques.

Critical to acid tailings reclamation protocol is the change in the physical structure of the wastes that permits improved infiltration and leaching. Addition of calcium compounds, fertilizer, and organic matter some time before seeding enhances infiltration and leaching. For continued plant growth, fertilization may be required for several years. Phosphorus is a problem since the increased pH of amended tailings permits the formation of insoluble metal/phosphorus complexes. Macronutrient and organic matter additions encourage the eventual reestablishment of nutrient cycles on the tailings. Properly functioning nutrient cycles ensure the development of a self-perpetuating, amendment-free plant community if the amended zone is deep enough to inhibit the diffusion of salts and acid into the new surface soil.

To encourage the germination and establishment of seeded species an improved seedbed is preferred to the tailings-lime-fertilizer-organic matter mix. The incorporation of compost or sewage sludge and industrial wastes such as fly ash with acid-generating materials to construct a seedbed has been demonstrated. The use of compost and fly ash as a soil substitute was successfully utilized in revegetation of denuded and eroded acid slopes in Pennsylvania in 1985.[39] The extension of this concept to acid tailings is appropriate.

Seedings on a tailings pond vary with climate and amendment. A clay- or limestone-capped tailings disposal area covered with a layer of coarse-textured overburden (up to 60 cm [2 ft]) and located in the northern Great Plains can be seeded with the species shown in Table 26. These plants emphasize tolerance to acid, salt, and drought. They are generally shallow rooted so that the toxic materials below the soil covering do not pose an insurmountable problem to plant survival. The shallow root system also encourages rapid water absorption in the near-surface soil horizon. This action inhibits acid formation deeper in the profile and salt migration into surface soil layers.

Of the plants selected for seeding (Table 26), slender wheatgrass is short lived but robust. It is a valuable nurse crop in this seeding. It provides protection for the slower-growing thickspike wheatgrass, mammoth wildrye *(Elymus giganteus)*, hard sheep fescue, pubescent wheatgrass *(Agropyron trichophorum)*, and birdsfoot trefoil *(Lotus corniculatus)*. Pubescent wheatgrass is coarse and not very palatable but produces well on sites with mild salt problems. As the plant community stabilizes, pubescent wheatgrass will disappear. Thus, the long-term plant community will probably be composed of thickspike wheatgrass, mammoth wildrye or hard sheep fescue, Canada bluegrass *(Poa compressa)*, remnants of the slender and pubescent wheatgrasses, any species volunteering from the surrounding area, and the two legumes from the original seed mix. These plants also constitute an adequate species representation for seeding on a tailings pond in which lime has been deeply incorporated. Germination and establishment are enhanced if the surface seed bed has been improved with lime, fertilizer, organic matter, and an inert material to dilute the tailings. The inert material may be any substance such as subsoil, fly ash, smelter slag, etc.

TABLE 26 Seed Mix for a Coarse-Textured Cover Soil on a Hypothetical Tailings Pond in the Semiarid West

Plant Lifeform	Common Name	Seed Rate[a] No./m²	Seed Rate[a] kg/ha	Seed Mix[b] (%)
Perennial				
Cool season grasses	Thickspike wheatgrass	50	1.5	15.5
	Mammoth wildrye	50	3.0	30.9
	Pubescent wheatgrass	50	2.6	26.8
	Canada bluegrass	80	0.2	2.1
	Hard sheep fescue	80	0.6	6.2
	Slender wheatgrass	30	1.0	10.3
Forbs				
	Yellow sweetclover	10	0.2	2.1
	Birdsfoot trefoil	50	0.6	6.2
		440	9.7	100.1

[a] PLS — No./m² ÷ 10.8 = No./ft²; kg/ha × 0.9 = lb/a.

[b] By weight.

Any plant species growing on tailings before reclamation should be included in the seed mix. Species living in the inhospitable environment of acid-generating wastes often have been genetically selected for tolerance to the adverse conditions of the site. They are adapted to the climatic, chemical, and physical characteristics of the wastes.

Populations of acid and metal-tolerant vegetation have been successfully selected and propagated for harsh site revegetation in Europe and Africa.[40] Unfortunately, these species require more moisture and milder climatic regimes than are available at most tailings disposal sites in North America. A selection of tolerant populations for revegetation of tailings on this continent would be very useful since these plants tolerate low nutrient concentrations as well as acid and high metal levels. Some tolerant ecotypes have been identified in North America but none has been propagated for seeding in tailings revegetation programs.[41,42]

The management program on tailings disposal sites should emphasize restricted grazing and other institutional controls. These sites will not produce vegetation in quantities comparable to growth rates on adjacent nonimpacted lands. Grazing must, therefore, be controlled to prevent overgrazing and erosion of the surface soil layer. Institutional controls on tailings ponds or other acid waste disposal areas should prohibit residential or industrial development.

WESTERN PHOSPHATE MINING

Western phosphate mining takes place in Idaho, Wyoming, Utah, and Montana, but the major portion of this activity is centered in southeastern Idaho on both private and National Forest administered lands. The ore is located in long, narrow, north-south-oriented mountain deposits (Figure 53). The area supports a valuable livestock grazing industry, a rich wildlife and fishery resource, unique aesthetic values, timber resources, and extensive watershed for the Columbia River. Legislation requiring reclamation of lands disturbed by this industry is very weak. Although revegetation of western phosphate mines is much easier and less expensive than reclamation in the semiarid coal fields or bentonite mining areas, it is not being pursued with equal vigor.

Until recently, phosphate ore removal was accomplished by a variation on the open-pit mining technique. Mining companies removed only the shallow-lying ore that was more amenable to treatment. They preferred to leave the pits open so that additional ore could be removed cheaply when the market improved at some future date. At these mines, the open pits remain as grim reminders of our abusive

Figure 53 Phosphate deposits of southeastern Idaho.

mining heritage. Overburden, and even topsoil, were excavated from these pits and dumped in head-of-hollow or valley fills rather than segregated and placed back into the pits. These waste dumps are one of the major reclamation problems associated with past mining in the phosphate area of southeastern Idaho.

Within the past 30 years, the postmining landscape was characterized by steep slopes with erosion channels, bare soils, or scattered clumps of grasses. Flat areas were often revegetated, but with introduced species. Few trees and shrubs were replacing the conifers and other woody plants destroyed by the mines. Livestock grazing and wildlife habitat were restricted to the few introduced grasses growing on flat surfaces.

More recent mining permits stress the needs of livestock and wildlife and include provisions for some backfilling of pits. Use of a modified open-pit technique has become common. Blocks or small open pits are constructed by separately removing topsoil, overburden, and ore. When the first cut is made, topsoil is typically placed in storage and overburden is dumped into a fill near the deposit. Topsoil from the second block is placed on the previously created topsoil storage pile, but overburden from this cut is backfilled into the first block. Topsoil from the third cut may be placed over the replaced fill in the first block and overburden into the second cut. This sequence may be continued indefinitely. The technique is relatively new in the region and, even when properly implemented, overburden from the new block is not adequate to completely fill the old cut; therefore, the depression created by removal of the ore is not completely backfilled. Approximate original contour cannot be attained with this practice alone.

Physical and chemical properties found in topsoils of this region are rarely inhibitory to plant growth and subsoils are generally non-toxic. Although surface soils may be deficient in nitrogen and potassium, phosphorus levels are usually adequate.[43] Precipitation is variable from dry valley bottoms to moist mountain slopes, but is adequate to support vegetation in all areas in most years. Winters are cold but not severe enough to hinder plant development. Despite the physical, chemical, and climatic amenities of this region, reclamation is not comparable to that found in the coal fields of the more inhospitable plains.

Reclamation in western phosphate fields is under the direction and regulation of the U.S. Forest Service or the state of Idaho. Major problems consist of:

1. Waste dump stabilization
2. Return to original contour
3. Poor species selection

Topsoil salvage is practiced at all phosphate mines today. This is probably the single most critical factor in the restoration of native vegetation on these disturbances. Studies showed that top dressings of at least 20 cm (8 in.) of this material consistently produced high-yielding grass stands in Idaho.[44] While the authors of these studies recommended fertilizer applications, research in western coal fields has shown that nitrogen, phosphorus, or potassium additions are not necessary when native plant species are seeded into direct-hauled topsoil.[45] The need for fertilizers on topsoiled phosphorus minesoils when seeded with all-native species should be investigated.

Topsoiled waste rock dumps are still characterized by erosion channels that would not be acceptable in the coal fields or at most metal mines of North America. This erosion contributes sediment to streams and loss of fishery productivity. Fortunately, geomorphic stability has not been a major problem of these waste rock dumps.

Slope reduction to facilitate equipment use and wildlife movement, to decrease soil erosion, and to increase aesthetic appeal and revegetation potential is needed. Slopes steeper than 2.5:1 are incompatible with good reclamation.[46] Good agronomic practices, including drill seeding, are impossible on such steep slopes because of the hazard posed to equipment. Seed and any amendments added to the soil are subject to loss by erosion or mass movement, resulting in bare hillsides. To permit the movement of equipment on the contour at these mines, and to facilitate seeding and soil amendment, slopes should be reduced to 3:1 or less.

Revegetation at western phosphate mines consists of seeding a few wheatgrasses, smooth and mountain brome, timothy, orchard grass *(Dactylis glomerata)*, and a few other minor species.[47,48] Seeding mixtures of native and introduced grasses is self defeating. Slower-growing native grasses are incompatible with aggressive introduced species. Waste dumps seeded to mixtures of native and naturalized species are dominated by the introduced grasses. Community succession and invasion of these sites by native vegetation, including tree species, are impeded and season-long grazing values are lowered by this dominance. Seed mixtures composed entirely of native species are now possible. Species selection should be based on predisturbance vegetation inventories and postmine land use. Releases of new grass and forb species by the Plant Materials Centers and commercial seed houses make native seed mixes possible and practical. Planting of browse composed of native shrubs and trees is also possible and increases wildlife and aesthetic values of the rehabilitated area.

The judicious incorporation of impoundments and bluff extensions in reclamation plans for these disturbances makes a major improvement in postmine landscape appearance and uses. Fill materials from depressions can be used to return other areas to approximate original contour. Permitting pond development in the depressions supplies a valuable livestock and wildlife water source as well as a potential fishery.

The present problems of waste dump erosion and introduced plant species need not be continued. The phosphate mining industry and the Forest Service have made progress during the past 20 years, but much remains to be accomplished. Perpetuation of past and some present mining practices are degrading our national forest heritage. This resource is increasing in value each day and it need not be damaged or destroyed by mining for phosphate ore. Reclamation plans that permit mining and preserve or enhance private and national forests can be written.

WESTERN BENTONITE MINING

Wyoming or sodium bentonite is a swelling clay mined for a large number of purposes.[49] It is found in several places in the West, but the best grades are mined in an arc around the western edge of the Black Hills of South Dakota and Wyoming. Sodium bentonite is composed of very fine particles, has a high cation exchange capacity, and is characterized by sodium dominance of the exchange sites.[50] It is mined in small pits that are usually less than 1 to 2 ha (2 to 5 acres) in size (Figure 54). It has been reported that bentonite beds in the northern Black Hills district range from 15 to 150 cm (0.5 to 5 ft) thick.[51] Mining disturbances, therefore, are characterized by numerous small, shallow pits scattered randomly over the landscape.

Revegetation problems associated with these materials and their environment are

1. Sodicity
2. Salinity
3. Fine texture
4. Poor soil structure contributing to lowered infiltration
5. Climatic extremes including drought and heat
6. Limited amount of topsoil

With the exception of sodium, trace element problems are not known to occur on these materials even though some are acid because of the oxidation of jarosite.

The chemical and physical characteristics of the materials found over, under, and around bentonite beds are similar to bentonite. They differ only by the degree of salinity, sodicity, and physical structure. Surface soils and overburden at these mines are usually heavy, blocky clays, slightly to strongly saline, and sodic. Such materials do not constitute the best surface soils for plant growth, but they are the

Figure 54 Bentonite deposit ready for removal from the pit.

only material available that is conducive to seed germination and plant establishment. Topsoil salvage becomes the dominant revegetation factor at these mines. Regardless of the condition of the native topsoil resource, it is the single most indispensable component of good revegetation of bentonitic minesoils. This surface layer is often very thin, and care must be exercised during stripping and placement to prevent contamination of the material by sodic spoil.

Abandoned bentonite pits and any low area created on landscapes mined for bentonite hold water. Bentonitic clays swell when wet and seal the bottom of the depressions, thus preventing infiltration. Water is lost only by overland flow or evaporation. These ponds often become suspensions of fine clay-sized particles, develop a skim of clay particles over the water, and are surrounded by saturated clays. The soft, muddy approaches to these impoundments make them death traps to any livestock or wildlife that attempt to use them as a water source. If ponded water is included in the reclamation plan for a bentonite mine, special provision must be made during the rehabilitation process to ensure that water quality is adequate for its intended use and that safe access to the water is provided for livestock and wildlife.

The bentonite mines west and north of the Black Hills are located in a hot dry area. Much of this region averages less than 40 cm (16 in.) of precipitation and some parts average less than 20 cm (8 in.). The revegetation program must include provisions for increasing water infiltration as well as incorporating drought- and salt-tolerant species in the seeding program. The limited precipitation that occurs must be held on site and not lost to runoff caused by the heavy clay texture of surface soils. The intense heat of the dark bentonite mine surface poses another problem for the revegetation specialist. Wood chips or straw mulches are an economical remedy for both infiltration and temperature problems. Mulches lower surface soil temperatures by shading and increase infiltration rates by providing pores into the heavy clay soils. Native hay mulches provide similar benefits and simultaneously provide a native plant seed source.

Vegetation selected for seeding on bentonite minesoils should emphasize tolerances to heavy clay, drought, salinity, and the shrink/swell characteristics of the soil. A seed mix for a hypothetical bentonite mine on the northwestern edge of the Black Hills is presented in Table 27. Plant communities in this mine area are characterized by the cool-season wheatgrasses, green needle grass, and needle-and-thread grass. Needle-and-thread grass is not available in commercial quantities and is rarely seeded in large revegetation programs. Bentonite minesoil revegetation has been successful when the remaining species are protected from grazing for three or more growing seasons. Unfortunately,

TABLE 27 Seed Mix for a Hypothetical Bentonite Mine in the Black Hills Mining Region

Plant Lifeform	Common Name	Seed Rate[a] No./m^2	Seed Rate[a] kg/ha	Seed Mix[b] (%)
Perennial				
Cool season grasses	Western wheatgrass	50	1.8	19.8
	Pubescent wheatgrass	50	2.6	28.6
	Green needlegrass	50	1.3	14.3
	Streambank wheat-grass	50	1.5	16.5
	Slender wheatgrass	30	0.9	9.9
Forbs				
	Yellow sweetclover	10	0.2	2.2
	Purple prairie clover	40	0.6	6.6
Shrubs				
	Gardner saltbush	5	0.2	2.2
		375	9.1	100.1

[a] PLS — No./m^2 ÷ 10.8 = No./ft^2; kg/ha × 0.9 = lb/a.

[b] By weight.

land ownership patterns make it difficult for mining companies to keep livestock off the newly seeded minesoils. Regulatory enforcement of permit restrictions is weak and encourages land owner violation of grazing limitations. Complete removal of revegetated lands from grazing for more than two growing seasons is not common.

Abandoned bentonite mines and their associated disturbances are widespread in the bentonite areas of Montana, Wyoming, and South Dakota. These sites have resisted revegetation attempts for many years due to high SARs, ECs, and expanding clay content. The state of Wyoming completed the reclamation of its abandoned bentonite minelands in 1990. The Wyoming reclamation program successfully utilized mulch rates up to 67 t/ha (30 t/a) of wood chips. Calcium compounds and fertilizer were also added to the abandoned spoils. These techniques were inferred from research that revealed a maximum production of native species at mulch rates of 90 t/ha (40 t/a) of woodchips.[52]

Other research has indicated that certain chemical treatments are capable of slowly changing the SAR (38 to 8) and EC (7.0 to 1.8 dS/m) of surface and near-surface abandoned bentonite minesoils.[53] These treatments increased infiltration and water storage, reduced soil surface crusting and bulk density, and improved soil structure. Changes were not immediately obvious but occurred over several years. Significant treatments included sulfuric acid with gypsum, calcium chloride with gypsum, magnesium chloride, and phosphogypsum. Six years after treatment grass production on some of these plots had increased from 416 kg/ha (370 lb/a) to over 2000 kg/ha (1780 lb/a) on plots treated with calcium chloride and gypsum. The longevity of these changes is still under investigation, but the increase in forage production on otherwise useless grazing land is remarkable.

SAND AND GRAVEL PITS

These sites are the most familiar land disturbance seen by urban residents in developed countries. The pits are numerous and found throughout the world. Sand and gravel mines are characterized by their long life. Although many mines consist of one small depression or excavation that is opened and then closed after a few months, most sand and gravel mines are active for several years. Some of these mines are active for decades or longer.[54] Because of the tendency to keep these mines open for many years, they serve as a focal point for weed infestation and weed seed distribution. Topsoil, spoil storage piles, and waste areas usually are not seeded and normally support only scattered pioneer plants. These plants are often noxious weeds.

Private and commercial sand and gravel mines are now controlled by permit restrictions in most states and provinces. The successful rehabilitation of these mines rests on close coordination and cooperation between regulators and miners. The main revegetation problems faced on the sites are

1. Limited financial resources
2. Coarse-textured materials and resulting poor water holding capacity, nutrient deficiencies, low CEC, and erosive materials
3. Lack of organic matter
4. Interception of watertables
5. Poor species selection

Because the vast majority of these mines are financially limited, the margin for error in revegetation programs is small. It is imperative that the rehabilitation program be completed successfully on the first attempt. It is the rare sand and gravel operator who can afford to return to a site several years after mineral extraction to correct mistakes or poor vegetation performance.

Topsoil salvage is required in permit stipulations by many state regulatory agencies. This action is the most important step in the successful revegetation of disturbances created by excavation for sand and gravel. With topsoil application and slope reduction, the drought problems inherent to sandy and gravelly soils are at least partially alleviated.

The nutrient deficiencies common to stored topsoil and to coarse-textured materials can be temporarily corrected with fertilizers, but the low CEC of sandy soils exacerbates this problem. Organic amendment helps reestablish nutrient cycles and increase the CEC, but the management necessary to maintain these accomplishments is not usually possible on small sand and gravel operations. All too often, livestock graze revegetated gravel pit slopes within one or two growing seasons after seed germination or before plant roots have developed enough to permit the plant to successfully withstand grazing.

These soils are very erosive. Organic matter, fertilizer, and rapidly growing vegetation are necessary to encourage soil stability and hold surface soils in place. The problems of drought and livestock access usually disrupt these goals on western sand or gravel pits.

Gravels are fluvially deposited and very permeable. Such deposits are often found near water bodies and hold near-surface aquifers. These water bodies may be pumped during the mining process but the cone of depression fills after mining. At many mines gravel extraction stops at the water line. At these mines the water table may be above ground during part of the year (Figure 55). This increases the difficulty of revegetation and the probability of pollution of the aquifer. To compound restoration problems these pits are commonly near urban areas and often have been used as landfills in the past. The indiscriminate dumping of household wastes into these pits has polluted many aquifers and surface water systems.

Species selected for revegetation at these mines are often based on recommendations developed for other types of mining and are not

Figure 55 Gravel pit with a high water table.

appropriate. Sandy soils require unique species selected for the root-zone materials and climate of the particular site. Several species that could be incorporated into a seed mix used on a sandy soil on the Great Plains are shown in Table 28. All of the plants in this table are native. The cool-season grasses selected for this site perform well on sandy, droughty soils. Streambank wheatgrass *(Agropyron riparium)* germinates, establishes, and grows rapidly on the site, but may gradually be displaced from the community by more aggressive rhizomatous grasses. Thickspike wheatgrass grows more slowly but persists and plants enlarge indefinitely. Needle-and-thread grass is a major component in mixed plant communities on most of the sandy soils of the plains. The extra expense of acquiring small quantities of seed of this species is balanced by its long-term persistence and contribution to soil stability. Prairie sandreed grass *(Calamovilfa longifolia)* is a sod former found in pure stands on sandy soils of the midgrass prairie. It is difficult to establish but can become a soil stabilizing species on even the most sandy soils of this region. Indian ricegrass germinates successfully in sandy soils of this region. It provides excellent long-term stability if other grasses do not crowd it out of the community.

Prairie clover is a native nitrogen fixer, a much needed component of the plant community for this site. Although yellow sweetclover *(Melilotus officinalis)* fixes greater quantities of nitrogen, it is not included in the seed mix because it readily volunteers on disturbances in this region. When present in high densities it inhibits grass

TABLE 28 Seed Mix for a Hypothetical Sand and Gravel Mine in the Northern Great Plains

		Seed Rate[a]		Seed Mix[b]
Plant Lifeform	Common Name	No./m^2	kg/ha	(%)
Perennial				
Cool season grasses	Thickspike wheatgrass	90	2.7	22.7
	Streambank wheatgrasses	60	4.3	36.1
	Needle-and-thread grass	60	1.9	16.0
Perennial				
Warm season grasses	Prairie sandreed	60	1.3	10.9
	Indian ricegrass	40	1.2	10.1
Forb				
	Purple prairieclover	30	0.5	4.2
		340	11.9	100.0
Shrub				
	Horizontal juniper	100/plants/ha		N.A.

[a] PLS — No./m^2 ÷ 10.8 = No./ft^2; kg/ha × 0.9 = lb/a.

[b] By weight — No./ha ÷ 2.5 = lb/a.

production. Seeding contributes to dominance of the site by this species. Horizontal juniper *(Juniperus horizontalis)* is difficult to establish from seed and requires transplanting of containerized or bareroot stock. It performs very well on loose, drifting sands in this region and makes a valuable contribution to the plant community on the revegetated sand and gravel pit.

As with bentonite revegetation, mining companies in dry rangeland environments usually cannot keep livestock off the site until vegetation has developed root systems that can withstand grazing. This is probably the single most deleterious factor contributing to accelerated erosion and revegetation failure on reclaimed sand and gravel mines.

HIGHWAY SHOULDERS

Diversity is not a characteristic of highway median or shoulder vegetation nor are multiple uses an intended attribute of these sites. Soil stabilization and aesthetically acceptable vistas are the only two uses for this vegetation.

The major obstacles to revegetation of highway rights-of-way are

1. Lack of topsoil
2. Steep slopes
3. Nutrient deficiencies
4. Poor species selection

All of these handicaps to successful revegetation are being addressed in some regions, but with varying degrees of emphasis.

Highway departments often consider topsoil as anything lying on the soil surface rather than the most desirable seed bed material. Plant response would be better if, before highway construction, surface soil were evaluated for revegetation purposes and less suitable surficial materials discarded. Poor quality surface soils are often used for the seedbed. Such soils require fertilization or other amendment. Fertilization stimulates plant growth and raises protein levels in plant tissues. More luxuriant growth and higher protein content encourages livestock and wildlife use of the shoulders and should be avoided if possible. Only limited macronutrient applications should be made along highway rights-of-way.

Highway cuts often have slopes steeper than 3:1. Such slopes encourage precipitation runoff and the ensuing aridity contributes to reduced seedling survival and plant growth. When steep slopes are combined with a south or west exposure and aridity they face an almost insurmountable handicap to plant survival. Erosion of the soil along the roadway is certain to follow.

Roadbeds have more dramatic changes in habitat than most other disturbances. They run through forest, rangeland, riparian habitat, urban zones, and agricultural land. They range from valley floor to mountain top. Nevertheless, road shoulders are usually seeded with generic mixes designed for average slope aspects, precipitation, elevation, and soil characteristics. Seed mixes for higher elevations, valley bottoms, and riparian areas rarely receive special consideration, despite the different ecological characteristics in these regions. Differing soil textures, salinity, or alkalinity also receive little or no consideration in highway shoulder seeding programs.

To some degree, one side of any road is exposed and the other protected. Along any stretch of highway two seed mixes are a minimum requirement, one mix for the exposed slope, the other for the more mesic, protected slope. Seed mixes should also change as road beds move from the valley floor to hilltop. The wide range of habitats and ecosystems impacted by the construction of highways and the varying degrees of site severity necessitate the development of a group of seed mixes for road shoulder revegetation.

The maintenance of highway rights-of-way is expensive. Grass cutting and weed control increase county or state highway expenses. Trees and large shrubs placed close to the roadway obscure the vision of drivers, encourage wildlife use, and require frequent trimming and cutting. Plant seed mixes selected for highway shoulders should be designed to reduce these costs.

Accidents between vehicles and wildlife or livestock are much too common along highways. They range from collisions between automobiles and large herbivores to swerving to avoid hitting birds or smaller animals. Collisions between vehicles and large herbivores increase hidden costs to local political bodies. Part of the responsibility for this expense and these accidents rests with highway departments that do not properly design road shoulder revegetation programs. Vegetation selected for seeding areas adjacent to highways or secondary roads should be chosen to discourage wildlife use. Seeding legumes and heavy fertilization encourages rather than discourages large and small mammals from using these areas. Rapidly growing large plant species offer protection and nesting sites to wildlife, thereby increasing the probability of vehicle-animal accidents. Plant species included in highway shoulder seed mixes should exhibit the following characteristics

1. Slow growth
2. Low growth form
3. Nonpalatable or relatively nonpalatable to large herbivores
4. Competitive
5. Adapted to the soils and climatic conditions of the site

Two seed mixes for shoulders along a highway at middle elevations in the northern Great Plains might include the species shown in Table 29. The three wheatgrasses and blue grama are very drought tolerant, an important characteristic for vegetation on south-facing slopes in this region. Two of the wheatgrasses are less than 60 cm (24 in.) tall and blue grama is a short grass, seldom growing over 15 to 20 cm (6 to 8 in.). Additional forbs might be added to close the canopy and slow or stop the entry of noxious weeds into the community.

On the north-facing hillslope the moisture deficit should be less severe than on the south-facing hillside; therefore, plants seeded on this hillside need not be as drought tolerant as the previous group. The three grasses selected for seeding are relatively short, generally less than 60 cm (2 ft) tall, and while grazed by livestock they are less palatable than many other species.

All highway shoulder revegetation programs should include noxious weed control. Because of slow and low growth of seeded vegetation, weeds present a problem along newly constructed shoulders for several years after seeding. Chemical control programs should not rely upon broadcast spraying of herbicides. Only the most selective

TABLE 29 Seed Mixes for an East-West Highway at Middle Elevations in the Northern Great Plains

Plant Lifeform	Common Name	Seed Rate[a] No./m²	Seed Rate[a] kg/ha	Seed Mix[b] (%)
	South Facing			
Perennial				
Cool season grasses	Streambank wheatgrass	70	2.1	27.3
	Crested wheatgrass	40	0.7	9.1
	Bluebunch wheatgrass	100	3.8	49.3
Perennial				
Warm season grass	Blue grama	80	0.5	6.5
Forb				
	Purple prairieclover	40	0.6	7.8
		330	7.7	100.0
	North Facing			
Perennial				
Cool season grasses	Western wheatgrass	70	2.8	70.0
	Hard sheep fescue	70	0.5	12.5
	Canada bluegrass	70	0.1	2.5
Forb				
	Purple prairieclover	40	0.6	15.0
		250	4.0	100.0

[a] PLS — No./m² ÷ 10.8 = No./ft²; kg/ha × 0.9 = lb/a.

[b] By weight.

use of chemical weed killers should be permitted. Indiscriminate application of herbicides retards the development of a balanced plant community that includes all life forms. The balanced plant community includes plants that deter or impede the invasion of the site by aggressive weedy plants. Booms spraying broadleaf herbicides along the entire width of the shoulder kill the forbs that occupy the niche otherwise utilized by noxious weeds. Rather than an asset to weed control, indiscriminate herbicide applications perpetuate a noxious weed control problem along highways. They preserve the highway right-of-way as an avenue of noxious weed dissemination by removing any competition for the weedy plants. Highway weed control programs based upon herbicides must emphasize spot application, not broadcast sprays.

ASH DISPOSAL PONDS

The expanding coal mining industry in North America is based on power generation and coal export. Metallurgical coal may bring a higher price, but more low-sulfur, thermal coals are being mined today than at any time in the recorded history of this continent. These coals range from 5 to 15% ash.[55,56] The ultimate fate of this very large quantity of ash poses a new revegetation problem.

Coal ash is composed of two fractions — bottom ash and fly ash. The concentrations of the major chemical components (e.g., aluminum, calcium, iron, magnesium, silicon, and sodium) in the two ash fractions are similar, but trace element concentrations (e.g., arsenic, boron, and lead) may differ by factors of 100 to 500 or more.[57] Relatively high boron and sodium contents are characteristic of thermal coals of the West. A number of analyses of coal ash are shown in Table 30.

TABLE 30 Chemical Constituents in Coal Ash[58,59]

Element	Range
Si	9–28%
Fe	2–24%
Al	1–21%
Ca	0.5–14%
Na	0.1–2.7%
Mg	0–2.4%
B	10–6500 μg/g
Pb	3–1600 μg/g
As	2–1700 μg/g
Se	1–500 μg/g

Ash from mine mouth power generating stations is occasionally disposed of in an open pit, strip, or underground coal mine. This disposal technique poses the problem of aquifer contamination with the salt or trace elements in the ash. Therefore, it is more common for both bottom and fly ash to be placed in ponds lined with membranes specifically designed for permanent ash disposal (Figure 56). Such holding ponds must be revegetated when filled.

The major plant-limiting factors associated with coal ash disposal sites are

1. Salinity
2. Elemental excesses or deficiencies including sodium and boron excess and macronutrient deficiencies
3. Organic matter deficiencies, and
4. Lack of soil structure

Specific sites may have unique problems but these four parameters represent the major plant growth-inhibiting characteristics of most ash disposal ponds.

These ponds are commonly covered with soil and seeded. If subsoil or mine overburden is used for the covering soil it must be amended with fertilizer and organic material. If topsoil was saved during the construction of the pond it may be spread over the subsoil layer.

Figure 56 Modern coal ash disposal pond with a heavy membrane liner.

Fertilization and organic matter needs are determined by the quality of the topsoil. To prevent the diffusion of salts from the ash zone into the overlying soil, the total depth of rootzone materials over toxic, saline coal ash should be at least 70 cm (28 in.) in semiarid western rangelands. Several studies have shown that this is a minimum depth to prevent salt diffusion into surface soils.[60,61] If the land is to be returned to forest or shrub growth, the depth of rootzone material should be increased. If the ash is not toxic, shallower coverings may be utilized.

Clay caps are placed over ash to prevent the diffusion of salt into surface soils as they are placed over tailings ponds to prevent the upward migration of acid and salts. Soil is placed over the clay cap to support the growth of plants. As much as 1 m (3.3 ft) of soil is often placed over the clay cap. This expense may not be necessary. Unless the ash is extremely toxic to plants, it need not be sealed from the plant rootzone. It is more important to ensure that salt does not migrate into the surficial soil layer than to seal the ash from the overlying soil.

Diffusion barriers have been used in reclamation programs to limit the diffusion of sodium and/or boron salts into coversoils. If a diffusion barrier is available to place over the ash to interrupt the flow of ions from deeper materials into the rootzone, the expense of the clay cap can be avoided. Plant roots readily grow through a diffusion barrier and penetrate into some underlying ashes. The phytotoxicity of the specific ash is the factor that controls root penetration.

In the seed mix selected for use on ash disposal ponds, shallow-rooted species should dominate (Table 31). Diversity requirements are rarely imposed on ash pond rehabilitation programs, therefore simple seed mixes (e.g., Table 31) are common on these sites. In future

TABLE 31 Seed Mix for a Hypothetical Ash Disposal Site on the Northern Great Plains

Plant Lifeform	Common Name	Seed Rate[a] No./m²	Seed Rate[a] kg/ha	Seed Mix[b] (%)
Perennial				
Cool season grasses	Western wheatgrass	80	2.8	32.9
	Slender wheatgrass	80	2.4	28.2
	Needle-and-thread grass	90	2.9	34.1
Forbs				
	Yellow sweetclover	5	0.1	1.2
	Purple prairieclover	20	0.3	3.5
		275	8.5	99.9

[a] PLS — No./m² ÷ 10.8 = No./ft²; kg/ha × 0.9 = lb/a.

[b] By weight.

use of the land, controls should prevent abuse of the vegetation resource and subsequent erosion of the soil covering.

Slender wheatgrass germinates very rapidly and grows vigorously. It provides all of the benefits of the nurse crop on this site. Western wheatgrass and needle-and-thread grass will be the dominant species on this pond, however, because slender wheatgrass will die out within a few growing seasons. Yellow sweetclover is seeded on this site because of its boron tolerance and nitrogen-fixing ability. Seed rates for this forb must be kept low, however, to prevent its dominance of the site. Purple prairie clover also fixes nitrogen but at a much slower rate. Because of the relatively small size of ash disposal ponds some invasion by native nonseeded species from adjoining areas may be anticipated. Complex seed mixes are not necessary on this size and type disturbance.

DRILL PADS

These sites are usually small and reclamation is not normally a problem. Their purpose and life varies, however, and older oil drill pads may be difficult to revegetate. Revegetation of these small sites consists of sealing the drill hole, ripping to destroy compaction that resulted from the movement of mechanized equipment on the pad, respreading topsoil (if it was salvaged prior to disturbance), seeding, and weed control. If the pad included a reservoir for drilling fluids, these materials must be neutralized or removed from the site prior to topsoiling and seeding. At almost all pads some amendments are necessary for adequate plant growth. At oil or gas pads that have been used for many years, fertilization and organic enrichment is required for adequate plant performance. The use of compost or sewage sludge should be considered on such sites, especially when topsoil is not available for application to the disturbance. Plant species selection should be based on the habitat, climate, soil conditions, and surrounding vegetation. Drill pads are small and surrounded by native vegetation. These species should be used in the revegetation program. Only a small number of species (two to four) need be seeded on these plots since they will be invaded by nonseeded species from the adjacent but undisturbed area.

ABANDONED MINE LANDS (AML)

Mining for coal, metals, and other minerals has left a legacy of abused and disturbed soils throughout the world. Because of the rich and widespread mineral deposits throughout North America, scattered areas on this continent have been laid waste by mining or mining related activities. These disturbances include both acid-generating and nonacid-producing waste rock, coal washings, open pits, overburden from surface and underground mines, as well as wastes from metal mining and processing facilities. Because of the diversity of materials and locations involved, it is impossible to suggest specific seed mixes for each AML site.

Over 337,000 ha (832,600 acres) of western land had been disturbed by surface mining prior to 1977.[62] One of the smaller but no less disruptive contributors to this disturbed land mass has been placer mining (Figure 57). This was one of the most destructive mining techniques, used to extract gold from stream-laid deposits. Many miles of stream channels were destroyed by this activity.

The spoil ridges of abandoned coal mines were a common phenomena as few as 20 years ago (Figure 58). Most of these surface scars have been corrected as part of the Abandoned Mined Lands (AML) program administered by the Office of Surface Mining. Unfortunately, metal mine and mill wastes remain highly visible throughout much of North America (Figure 59). A major political dispute has arisen over the definition of many of these metal mine waste sites in the United States. The dispute calls into question their designation as simple AMLs. Some public interest groups and federal agencies believe that the metalliferous wastes at many of these disturbances should not be considered AMLs, but rather classified as hazardous waste sites. The added costs associated with labelling degraded areas as

Figure 57 Abandoned placer mine waste.

Figure 58 Abandoned coal minespoils (early 1970).

Figure 59 Abandoned wastes from hardrock mining and milling.

hazardous waste sites have stopped or hindered cleanup of many of these mining-related waste deposits.

Acid generation is characteristic of many of these waste sites. Such materials are especially difficult to reclaim. They present the problems noted for acid-generating materials in the discussion of *Metal Mine*

Wastes earlier in this chapter. Small deposits of abandoned, acid-generating mining wastes should be excavated if they present a potential pollutant source to surface or ground waters. When these wastes are within flood plains, they should be pulled back from the stream so that they neither influence nor are influenced by the stream. Permanent placement of these materials should be above any watertable and they should be revegetated and protected from erosion.

Growth of vegetation on AML sites is controlled by the characteristics of the rooting material and the climate of the site. The major handicap to revegetation is the absence of a surficial seedbed. In the semiarid West the surface soil is thin and large quantities of topsoil are rarely available for salvage or borrowing. Even in the more mesic eastern part of the country, topsoil is rarely available in quantities adequate to cover large AML disturbances. Therefore, it is often necessary to plan revegetation programs on these sites without a surface soil covering. The use of compost and/or sewage sludge to construct a surface rooting and seed germinating layer is receiving increased use on these types of wastes.[63,64]

In general, revegetation of AML sites should involve seeding of species that are adapted to the climate, rootzone materials, and exposure of the site. The interested reader may consult the previous sections of this chapter for suggestions on seed mixes for specific materials and locations.

REFERENCES

1. Soil Conservation Service, Plant Materials Centers, Plant Materials Handbook, Bureau of Mines, U.S. Department of the Interior, Denver, CO, 1988.
2. Hardy BBT Limited, *Manual of Plant Species Suitability for Reclamation in Alberta*, 2nd ed., Alta. Land Conserv. Reclam. Coun. Rep. No. RRTAC 89-4, Calgary, 1989.
3. Munshower, F. F., *Perennial Grasses for Revegetation of Disturbed Lands in the Northern Great Plains and the Intermountain Region*, 2nd ed., Reclam. Res. Unit Publ. No. 9101, Montana State University, Bozeman, 1991.
4. Munshower, F. F., *Forbs, Shrubs, and Trees for Revegetation of Disturbed Lands in the Northern Great Plains and Adjacent Areas*, Reclam. Res. Unit Publ. No. 9102, Montana State University, Bozeman, 1991.
5. Soil Survey Staff, *Soil Survey Manual*, Handbook No. 18., U.S. Department of Agriculture, Washington, D.C., 1951.
6. Council on Environmental Quality, Regulation of Surface Mining and Reclamation for Minerals Other Than Coal, A report to the President and Congress pursuant to Sect. 709 of SMCRA 1977, U.S. Department of the Interior, Washington, D.C., 1981, 21.

7. Spry, M. J., Revegetation Research on Hard Rock Mining Disturbances in North-Central Montana, M.S. thesis, Montana State University, Bozeman, 1986, 58.
8. Sopper, W. E., Reforestation of a zinc smelter superfund site, in Mine Drainage and Surface Mine Reclamation Vol. II, Bur. of Mines Inf. Circ. 9184, U.S. Department of the Interior, Pittsburgh, PA, 1988, 68.
9. Vohdenal, G. L., The use of municipal compost in the revegetation of a high elevation mine site, in *Proc. Billings Mined Land Reclam. Symp.*, Reclam. Res. Unit Publ. No. 9301, Montana State University, Bozeman, 1993, 30.
10. Mueggler, W. F. and Stewart, W. L., Grassland and Shrubland Habitat Types of Western Mont, Gen. Tech. Rep. INT-66, Interm. For. Range Exp. Sta., U.S. Department of Agriculture, Ogden, UT, 1980.
11. Doll, E. C., Merrill, S. D., and Halvorson, G. A., *Soil Replacement for Reclamation of Stripmined Lands in North Dakota,* Agric. Exp. Sta. Bull. 514, North Dakota State University, Fargo, 1984, 5.
12. Power, J. F., Sandoval, F. M., and Ries, R. E., Topsoil-subsoil requirements to restore North Dakota mined lands to original productivity, *Min. Eng.*, Dec., 1708, 1979.
13. Brown, R. W., Johnson, R. S., Richardson, B. Z., and Farmer, E. E., Rehabilitation of alpine disturbances: Beartooth Plateau, Montana, in *Proc. High Altitude Reveg. Workshop No. 2,* Zuck, R. H. and Brown, L. F., Eds., Environmental Resources Center, Colorado State University, Fort Collins, 1976, 58.
14. Lawson, J. A., Suitability of an Alluvial Overburden Material as a Plant Growth Medium at the Berkeley Complex in Butte, Montana, M.S. thesis, Montana State University, Bozeman, 1984.
15. Spry, M. J., Revegetation Research on Hard Rock Mining Disturbances in North-Central Montana, M.S. thesis, Montana State University, Bozeman, 1986.
16. Williams, E. D., Rehabilitation of Fire Suppression Impacts on the North Fork Fire in Yellowstone National Park, Paper presented Meet. Am. Soc. Surf. Min. Reclam., Drango, CO, 1991.
17. Williamson, N. A., Johnson, M. S., and Bradshaw, A. D., *Mine Waste Reclamation,* Mining Journal Books Ltd., London, 1982, 1.
18. Williamson, N. A., Johnson, M. S., and Bradshaw, A. D., *Mine Waste Reclamation,* Mining Journal Books Ltd., London, 1982, 27.
19. Nielson, R. F. and Peterson, H. B., *Treatment of Mine Tailings to Promote Vegetative Stabilization,* Utah Agric. Exp. Sta. Bull. 485, Utah State University, Logan, 1972.
20. Nielson, R. F. and Peterson, H. B., Ut. Agric. Exp. Sta. Bull. 485, Utah State University, Logan, 1972.
21. Williamson, N. A., Johnson, M. S., and Bradshaw, A. D., *Mine Waste Reclamation,* Mining Journal Books Ltd., London, 1982, 27.
22. Caruccio, F. T., Hossner, L. R., and Geidel, G., Pyritic materials: acid drainage, soil acidity, and liming, in *Reclamation of Surface-Mined Lands,* Vol. 1, CRC Press, Boca Raton, FL, 1988, 159.

23. Sobek, A. A., Hossner, L. R., Sorensen, D. L., Sullivan, P. J., and Fransway, D. F., Acid-base potential and sulfur forms, in *Reclaiming Mine Soils and Overburden in the Western United States,* Williams, R. D. and Schuman, G. E., Eds., Soil Conservation Society of America, Ankeny, IO, 1987, 233.
24. Dean, K. C., Froisland, L. J., and Shirts, M. B., Utilization and Stabilization of Mineral Wastes, Bur. of Mines Bull. 688, U.S. Department of the Interior, Washington, D.C., 1986, 16.
25. Ritcey, G. M., *Tailings Management,* Elsevier, New York, 1989, 753.
26. Johnson, M. S. and Bradshaw, A. D., Ecological principles for the restoration of disturbed and degraded lands, *Appl. Biol.,* 4, 141, 1979.
27. Seidel, K. W., Tolerance of seedlings of Ponderosa pine, Douglas fir, Grand fir, and Engelman spruce for high temperatures, *Northwest Sci.,* 60, 1, 1986.
28. Barth, R. C. and Martin, B. K., *Soil-Depth Requirements to Reclaim Surface-Mined Areas in the Northern Great Plains,* Colorado School of Mines, Golden, 1982, 42.
29. Power, J. F., Sandoval, F. M., and Ries, R. E., Topsoil-subsoil requirements to restore North Dakota mined land to original productivity, *Min. Eng.,* Dec., 1708, 1979.
30. Barth, R. C., *Soil-Depth Requirements to Reestablish Perennial Grasses on Surface-Mined Areas in the Northern Great Plains,* Miner. Energ. Resour. 27, Colorado School of Mines, Golden, 1984.
31. Craze, B., Restoration of Captains Flat mining area, *J. Soil Conserv. Serv. N.S.W.,* 33, 98, 1977.
32. Keammerer, W. R., Arthur, D., and Kuenstling, A., *Anaconda Longterm Vegetation Monitoring Project 1988–1990,* Rep. prep. Keammerer Ecological Consultants, Inc. for ARCO, Anaconda, MT, 1992, 721 pp.
33. Cedergren, H. R., *Seepage, Drainage, and Flow Nets,* 3rd ed., John Wiley & Sons, New York, 1989.
34. Nielson, R. F. and Peterson, H. B., *Treatment of Mine Tailings to Promote Vegetative Stabilization,* Utah Agric. Exp. Sta. Bull. 485, Utah State University, Logan, 1972, 16.
35. Hunter, G. D. and Whiteman, P. C., Revegetation of mine wastes at Mt. Isa, Queensland. II. Amendment of nutrient status and physical properties of tailings for plant growth, *Aust. J. Exp. Agric. Anim. Husb.,* 15, 803, 1975.
36. Hill, R. D., Hinkle, K. R., and Apel, M. L., Reclamation of Pyritic Waste, Publ. EPA/600/14, Off. Res. Dev., U.S. Environmental Protection Agency, Cincinnati, OH, 1982.
37. Schafer and Associates and Reclamation Research Unit, *Stream Bank Tailings and Revegetation Studies, STARS Phase II: Field Scale Treatability Study Plot Construction Final Summary Report,* Montana Department of Health and Environmental Science, Helena, 1989.
38. Dollhopf, D. J., *Deep Lime Incorporation Methods for Neutralization of Acidic Minesoils,* Reclam. Res. Unit Publ. 9201, Montana State University, Bozeman, 1992.

39. Oyler, J. A., Reclamation of a site near a smelter using sludge/fly ash amendments: herbaceous species, in Mine Drainage and Surf. Mine Reclam. Proc. Symp. Vol. II, Bur. of Mines Inf. Circ. 9184, U.S. Department of the Interior, Pittsburgh, PA, 1988, 22.
40. Smith, R. A. and Bradshaw, A. D., Stabilization of toxic mine wastes by the use of tolerant plant populations, *Trans. Inst. Min. Metall.*, 81, A230, 1972.
41. Surbrugg, J. E., Copper and Zinc Tolerance in Two Montana Grass Species Growing on Copper Mill Tailings, M.S. thesis, Montana State University, Bozeman, 1982, 54.
42. Cox, R. M. and Hutchinson, T. C., Multiple metal tolerances in the grass *Deschampsia cespitosa* (L.) Beauv. from the Sudbury smelting area, *New Phytol.*, 84, 631, 1980.
43. Farmer, E. E. and Blue, W. G., Reclamation of lands mined for phosphate, in *Reclamation of Drastically Disturbed Lands*, Schaller, F. W. and Sutton, P., Eds., American Society of Agronomy, Madison, WI, 1978, 585.
44. Farmer, E. E. and Blue, W. G., Reclamation of Drastically Disturbed Lands, Schaller, F. W. and Sutton, P., Eds., American Society of Agronomy, Madison, WI, 1978, 585.
45. Rennick, R. B., Hertzog P. J., and Munshower, F. F., *Native Species Response to Fertilizers on Surface Mined Land*, Agric. Exp. Sta. Spec. Rep. 11, Montana State University, Bozeman, 1984, 60.
46. Farmer, E. E. and Blue, W. G., Reclamation of lands mined for phosphate, in *Reclamation of Drastically Disturbed Lands*, Schaller, F. W. and Sutton, P., Eds., American Society of Agronomy, Madison, WI, 1978, 585.
47. Farmer, E. E. and Richardson, B. Z., Phosphate mine dump stability and revegetation, in *Proc. Symp. Watershed Management V.*, II., American Society of Civil Engineers, New York, 1980, 888.
48. Richardson, B. Z. and Farmer, E. E., Revegetation of phosphate mined lands in the intermountain west, in *Proc. Symp. Reclam. Phosphate Ind.*, Florida Institute of Phosphate Research, Bartow, 1983, 373.
49. Daily Post, *Bentonite, The Clay of 1,000 Uses*, Belle Fourche, SD, p. 5, Sept. 26, 1972.
50. Bjugstad, A. J., Bentonite mine spoil and pit reclamation, in Proc. Miner. Waste Stabilization Liaison Comm., Evelith, MN, 1979.
51. Knechtel, M. M. and Patterson, S. H., Bentonite Deposits of the Northern Black Hills District, Wyoming, Montana and South Dakota, Bull. 1082-M, U.S. Geological Survey, Denver, CO, 1962, 893.
52. Smith, J. A., DePuit, E. J., and Schuman, G. E., Wood residue and fertilizer amendment on bentonite mine spoils. II. Plant species response, *J. Environ. Qual.*, 15, 427, 1986.
53. Dollhopf, D. J., Rennick, R. B., and Smith, S. C., *Long-Term Effects of Physiological Amendments on Plant Performance at a Bentonite Minesite in the Northern Great Plains*, Reclam. Res. Unit Publ. No. 8802, Montana State University, Bozeman, 1988, 3.

54. Blauch, B. W., Reclamation of lands disturbed by stone quarries, sand and gravel pits, and borrow pits, in *Reclamation of Drastically Disturbed Lands*, Schaller, F. W. and Sutton, P., Eds., American Society of Agronomy, Madison, WI, 1978, 619.
55. Tripodi, R. A. and Cheremisinoff, P. N., *Coal Ash Disposal: Solid Waste Impacts*, Technomic Publ., Westport, CT, 1980, 12.
56. Tripodi, R. A. and Cheremisinoff, P. N., *Coal Ash Disposal: Solid Waste Impacts*, Technomic Publ., Westport, CT, 1980, 18.
57. Cooley, S. A. and Ellmen, R. C., Analysis of the Northern Great Plains Province Lignites and Subbituminous Coals and Their Ash, Publ. DOE/GFETC/RI-81/2, U.S. Department of Energy, Grand Forks, ND, 1981, 15.
58. Wewerka, E. M., Williams, J. M., Wanek, P. L., and Olsen, J. D., Environmental Contamination from Trace Elements in Coal Preparation Wastes, EPA-600/7–76–007, NTIS No. PB 267 339, U.S. Environmental Protection Agency, Research Triangle Park, NC, 1976, 13.
59. Tripodi, R. A. and Cheremisinoff, P. N., *Coal Ash Disposal: Solid Waste Impacts*, Technomic Publ., Westport, CT, 1980, 18.
60. Power, J. F., Sandoval, F. M., and Ries, R. E., Topsoil-subsoil requirements to restore North Dakota mined lands to original productivity, *Min. Eng.*, Dec., 1708, 1979.
61. Barth, R. C. and Martin, B. K., *Soil-Depth Requirements to Reclaim Surface-mined Areas in the Northern Great Plains*, Colorado School of Mines, Golden, 1982, 11.
62. Soil Conservation Service, The Status of Land Disturbed by Surface Mining in the United States, S.C.S.-TP-158, U.S. Department of Agriculture, Washington, D.C., 1979.
63. Brandt, C. A. and Hendrickson, P. L., *Use of Composts in Revegetating Arid Lands*, PNL—7833, Battelle Pacific Northwest Laboratory, Richland, WA, 1991, 31.
64. Oyler, J. A., Reclamation of site near a smelter using sludge/fly ash amendments: herbaceous species, in Mine Drainage Surf. Mine Reclam. Proc. Symp. Vol. II, Bur. of Mines Inf. Circ. 9184, U.S. Department of the Interior, Pittsburgh, PA, 1988, 22.

APPENDIX

Genus and Species of Plants Commonly Used in Rehabilitation Programs

TABLE 32 Plant Species Used in Eastern Revegetation Programs

Grasses	
Scientific Name[a]	Common Name
Agrostis alba L.	Redtop
A. tenuis Sibth.	Colonial bentgrass
Andropogon gerardi Vitman	Big bluestem
Arrhenatherum elatius (L.) Presl.	Tall oatgrass
Bothriochloa caucasica (Trin.) Hubb.[b]	Caucasian bluestem
Bouteloua curtipendula (Michx.) Torr.	Sideoats grama
B. gracilis (H.B.K.) Lag. ex Steud.	Blue grama
Bromus arvensis L.	Field brome
B. inermis Leyss.	Smooth brome
Buchloe dactyloides (Nutt.) Engelm.	Buffalograss
Cynodon dactylon (L.) Pers.	Bermuda grass
Dactylis glomerata L.	Orchard grass
Deschampsia caespitosa (L.) Beauv.	Tufted hairgrass
Echinochloa crusgalli (L.) Beauv.	Barnyard grass
E. crusgalli var. *frumentacea* (Link) W.F. Wright	Japanese millett
Eragrostis curvula (Schrad.) Nees	Weeping lovegrass
Festuca arundinacea Schreb.	Kentucky "31" tall fescue
F. rubra L.	Red fescue
Lolium multiflorum Lam.	Italian ryegrass
L. perenne L.	Perennial ryegrass
Panicum clandestinum L.	Deertongue
P. miliaceum L.	Proso millet
P. virgatum L.	Switchgrass
Paspalum dilatatum Poir.	Dallis grass
P. notatum Flugge	Bahiagrass
Pennisetum americanum (L.) K. Schum.[b]	Pearl millet
Phalaris arundinacea L.	Reed canarygrass
Phleum pratense L.	Timothy

Poa compressa L.	Canada bluegrass
P. pratensis L.	Kentucky bluegrass
Schizachyrium scoparium (Michx.) Nash[b] (*Andropogon scoparius* Michx. of Hitchcock 1950)	Little bluestem
Setaria italica (L.) Beauv.	Foxtail millet
Sorghastrum nutans (L.) Nash	Indiangrass
Sorghum sudanense (Piper) Stapf.	Sudangrass
S. vulgare Pers. [including *S. bicolor* (L.) Moench]	Sorghum

Forbs	
Scientific Name[c]	Common Name
Astragalus cicer L.[d]	Cicer milkvetch
Cassia fasciculata Michx.	Partridge pea
Coronilla varia L.	Crownvetch
Dalea purpurea Vent.	Purple prairie clover
Desmanthus illinoensis (Michx.) MacM.	Illinois bundleflower
Fagopyrum esculentum Moench	Buckwheat
F. sagittatum Gilib.	Buckwheat
Helianthus annuus L.	Annual sunflower
H. maximiliani Schrad.	Maximilian sunflower
Lathyrus sylvestris L.[b]	Flatpea
Lespedeza cuneata (Dum.Cours.) G.Don	Sericea lespedeza
L. daurica (Laxm.) Schindl.	Prostrate lespedeza
L. stipulacea Maxim.[b]	Korean lespedeza
L. striata (Thunb.) Hook. & Arn.[b]	Common and Kobe lespedeza
Liatris pycnostachya Michx.	Prairie blazing star
Lotus corniculatus L.	Birdsfoot trefoil
L. tenuis Waldst. & Kit. ex Willd.	Narrowleaf trefoil
Medicago sativa L.	Alfalfa
Melilotus alba Medic.	White sweetclover
M. officinalis (L.) Lam.	Yellow sweetclover
Onobrychis viciaefolia Scop.	Sainfoin
Penstemon spp. Mitch.	Beardstongue
Polygonum cuspidatum Sieb. & Zucc.	Japanese fleeceflower
Ratibida columnifera (Nutt.) Woot. & Standl.	Prairie coneflower
Salvia azurea Michx. & Lam.	Blue sage
Trifolium hybridum L.	Alsike clover
T. incarnatum L.	Crimson clover
T. pratense L.	Red clover
T. repens L.	Ladino or white clover
T. subterraneum L.[b]	Subterranean clover
Vicia sativa L.	Common vetch
V. villosa Roth	Hairy vetch
Vigna unguiculata (L.) Walp.	Black-eyed cowpea

Shrubs	
Scientific Name[c]	Common Name
Amorpha fruticosa L.	Indigobush
Caragana arborescens Lam.	Siberian peashrub
Cornus amomum Mill.	Silky dogwood
C. racemosa Lam.	Gray dogwood
C. stolonifera Michx.	Redosier dogwood
Crataegus spp. L	Hawthorne
Elaeagnus umbellata Thunb.	Autumn olive
Lespedeza bicolor Turcz.	Shrub lespedeza
L. japonica Bailey[b]	Japanese lespedeza
L. thunbergii (DC.) Nakai	Thunberg lespedeza
Ligustrum amurense Carriere[b]	Amur privet
Lonicera japonica Thunb.	Japanese honeysuckle
L. maackii (Rupr.) Maxim.	Amur honeysuckle
L. morrowi Gray	Morrow honeysuckle
L. tatarica L.	Tatarian honeysuckle
Prunus americana Marsh.	American plum
P. besseyi L.H. Bailey[b]	Western sandcherry
P. virginiana L.	Chokecherry
Rhus aromatica Ait.	Fragrant sumac
R. copallina L.	Shining sumac
Robinia fertilis Ashe[b]	Locust
R. hispida L.d[b]	Bristly locust
Rosa rugosa Thunb.	Rugosa rose
Salix purpurea L.	Purple osier willow
Sambucus canadensis L.	American elder
Shepherdia argentea (Pursh) Nutt.[b]	Silver buffaloberry
Vaccinium L.	Blueberry
Viburnum dentatum L.	Southern arrowood
V. opulus L.	High-bush cranberry

Trees	
Scientific Name[c]	Common Name
Acer rubrum L.	Red maple
A. saccharinum L.	Silver maple
A. saccharum Marsh.	Sugar maple
Alnus glutinosa (L.) Gaertn.	European black alder
Betula lenta L.[b]	Sweet birch
B. nigra L.	River birch
B. papyrifera Marsh.	Paper birch
B. pendula Roth.[b]	European white birch
B. populifolia Marsh.	Gray birch
Carya spp. Nutt.	Hickory
Carya illinoensis (Wang.) Koch	Pecan
Castanea mollissima Bl.	Chinese chestnut
Catalpa speciosa (Warder) Englem.	Northern catalpa

Celtis occidentalis L.	Hackberry
Cornus florida L.	Flowering dogwood
Elaeagnus angustifolia L.	Russian olive
Fraxinus americana L.	White ash
F. pennsylvanica Marsh.	Green ash
Juglans nigra L.	Black walnut
Juniperus virginiana L.	Eastern red cedar
Larix decidua Mill.	European larch
L. leptolepis (Sieb. & Zucc.) Gord.[e]	Japanese larch
Liquidambar styraciflua L.	Sweetgum
Liriodendron tulipifera L.	Yellow poplar
Maclura pomifera (Raf.) Schneid.	Osage orange
Malus Mill.	Crab apple
Morus rubra L.	Mulberry
Paulownia tomentosa (Thunb.) Sieb. & Zucc.	Royal paulownia
Picea abies (L.) Karst	Norway spruce
P. glauca (Moench) Voss[b]	White spruce
Pinus banksiana Lamb.	Jack pine
P. echinata Mill.	Shortleaf pine
P. elliottii Engelm.[b]	Slash pine
P. nigra Arnold	Austrian pine
P. palustria Mill.	Longleaf pine
P. resinosa Ait.	Red pine
P. rigida Mill.	Pitch pine
P. strobus L.	Easternwhite pine
P. sylvestris L.	Scotch pine
P. taeda L.	Loblolly pine
P. virginiana Mill.[b]	Virginia pine
Platanus occidentalis L.	American sycamore
Populus deltoides Marsh.	Eastern cottonwood
Prunus serotina Ehrh.	Black cherry
Pseudotsuga menziesii (Mirb.) Franco[b]	Douglas fir
Quercus acutissima Carruth.[b]	Sawtooth oak
Q. alba L.	White oak
Q. imbricaria Michx.	Shingle oak
Q. macrocarpa Michx.	Bur oak
Q. palustris Muenchh.	Pin oak
Q. rubra L.	Northern red oak
Robinia pseudoacacia L.	Black locust
Salix spp. L.	Willow
S. nigra Marsh.	Black willow
Taxodium distichum (L.) Rich.	Bald cypress
Tilia americana L.	Basswood

[a] Nomenclature from Hitchcock, 1950 except as noted.
[b] Nomenclature from Bailey and Bailey, 1976.
[c] Nomenclature from Mohlenbrock, 1986 except as noted.
[d] Nomenclature from Hitchcock and Cronquist, 1973.
[e] Nomenclature from Rehder, 1954.

TABLE 33 Plant Species Used in Midwestern Revegetation Programs

Grasses

Scientific Name[a]	Common Name
Agrostis alba L.	Redtop
Andropogon gerardi Vitman	Big bluestem
Bromus inermis Leyss.	Smooth brome
Cynodon dactylon (L.) Pers.	Bermudagrass
Dactylis glomerata L.	Orchardgrass
Deschampsia caespitose (L.) Beauv.	Tufted hairgrass
Echinochloa crusgalli (L.) Beauv.	Barnyard grass
E. crusgalli var. *frumentacea* (Link) W.F.Wright	Japanese millet
Eragrostis curvula (Schrad.) Nees	Weeping lovegrass
E. trichodes (Nutt.) Wood	Sand lovegrass
Festuca arundinacea Schreb.	Kentucky "31" Tall fescue
F. rubra L.	Red fescue
Lolium multiflorum Lam.	Italian ryegrass
L. perenne L.	Perennial ryegrass
Panicum clandestinum L.	Deertongue
P. virgatum L.	Switchgrass
Pennisetum americanum (L.) K.Schum[b]	Pearl millet
Phalaris arundinacea L.	Reed canarygrass
Phleum pratense L.	Timothy
Poa compressa L.	Canada bluegrass
P. pratensis L.	Kentucky bluegrass
Schizachyrium scoparium (Michx.) Nash[b] (*Andropogon scoparius* Michx. of Hitchcock 1950)	Little bluestem
Setaria italica (L.) Beauv.	Foxtail millet
Sorghastrum nutans (L.) Nash	Indiangrass
Sorghum sudanense (Piper) Stapf.	Sudangrass
S. vulgar Pers. including *S. bicolor* (L.) Moech.	Sorghum

Forbs

Scientific Names[c]	Common Names
Amorpha canescens Pursh	Leadplant
Astragalus cicer L.[d]	Cicer milkvetch
Coronilla varia L.	Crownvetch
Helianthus annuus L.	Annual sunflower
H. maximiliani Schrad.	Maximilian sunflower
Lespedeza cuneata (Dum.Cours.) G. Don	Sericea lespedeza
L. stipulacea Maxim[b]	Korean lespedeza
L. striata (Thunb.) Hook. & Arn.[b]	Common and Kobe lespedeza
Lotus corniculatus L.	Birdsfoot trefoil
Medicago sativa L.	Alfalfa
Melilotus alba Medic.	White sweetclover

M. officinalis (L.) Lam.	Yellow sweetclover
Onobrychis viciaefolia Scop.	Sainfoin
Trifolium hybridum L.	Alsike clover
T. pratense L.	Red clover
T. repens L.	Ladino or white clover
Vicia villosa Roth	Hairy vetch

Shrubs

Scientific Name[c]	Common Name
Caragana arborescens Lam.	Siberian peashrub
Cornus amomun Mill	Silky dogwood
C. racemosa Lam.	Gray dogwood
C. stolonifera Michx.	Red-osier dogwood
Crataegus spp. L.	Hawthorn
Elaeagnus umbellata Thunb.	Autumn olive
Lespedeza bicolor Turcz.	Shrub lespedeza
Rhus copallina L.	Shining sumac
Robinia hispida L.	Bristly locust

Trees

Scientific Name[c]	Common Name
Acer rubrum L.	Red maple
A. saccharinum L.	Silver maple
A. saccharum Marsh.	Sugar maple
Ailanthus altissima (Mill.) Swingle	Tree-of-heaven
Alnus glutinosa (L.) Gaertn.	European black alder
Betula nigra L.	River birch
Carya spp. Nutt.	Hickory
C. illinoensis (Wang.) Koch	Pecan
Celtis occidentalis L.	Hackberry
Cornus florida L.	Flowering dogwood
Elaeagnus angustifolia L.	Russian olive
Fraximus americana L.	White ash
F. pennsylvanica Marsh.	Green ash
Juglans nigra L.	Black walnut
Juniperus virginiana L.	Eastern red cedar
Liquidambar styraciflua L.	Sweetgum
Liriodendron tulipifera L.	Yellow poplar
Maclura pomifera (Raf.) Schneid.	Osage-orange
Pinus banksiana Lamb.	Jack pine
P. echinata Mill.	Shortleaf pine
P. nigra Arnold	Austrian pine
P. resinosa Ait.	Red pine

P. rigida Mill.	Pitch pine
P. strobus L.	Eastern white pine
P. sylvestris L.	Scotch pine
P. taeda L.	Loblolly pine
P. virginiana Mill.[b]	Virginia pine
Platanus occidentalis L.	American sycamore
Populus deltoides Marsh.	Eastern cottonwood
Prunus serotina Ehrh.	Black cherry
Quercus acutissima Carruth.[b]	Sawtooth oak
Q. alba L.	White oak
Q. imbricaria Michx.	Shingle oak
Q. macrocarpa Michx.	Bur oak
Q. pallustris Muenchh.	Pin oak
Q. prinus L.	Chestnut oak
Q. rubra L.	Northern red oak
Robinia pseudoacacia L.	Black locust
Taxodium distichum (L.) Rich.	Bald cypress

[a] Nomenclature from Hitchcock, 1950 except as noted.
[b] Nomenclature from Bailey and Bailey, 1976.
[c] Nomenclature from Mohlenbrock, 1986 except as noted.
[d] Nomenclature from Hitchcock and Cronquist, 1973.

TABLE 34 Plant Species Used in Northern Great Plains Revegetation Programs

Grasses	
Scientific Name[a]	Common Name
Agropyron cristatum (L.) Gaertn.	Crested wheatgrass
A. dasystachyum (Hook.) Scribn.	Thickspike wheatgrass
A. desertorum (Fisch.) Schult.	Desert wheatgrass
A. elongatum (Host) Beauv.[b]	Tall wheatgrass
A. inerme (Scribn. & Smith) Rydb.	Beardless wheatgrass
A. intermedium (Host) Beauv.	Intermediate wheatgrass
A. riparium Scribn. & Smith	Streambank wheatgrass
A. sibiricum (Willd.) Beauv.	Siberian wheatgrass
A. smithii Rydb.	Western wheatgrass
A. spicatum (Pursh) Scribn. & Smith	Bluebunch wheatgrass
A. trachycaulum (Link) Malte	Slender wheatgrass
A. trichophorum (Link) Richt.	Pubescent wheatgrass
Agrostis alba L.	Red top
A. scabra Willd.	Hair grass, ticklegrass
A. stolonifera L.	Creeping bentgrass
A. tenuis Sibth.	Colonial bentgrass
Alopecurus arundinaceus Poir.	Creeping foxtail
A. pratensis L.	Meadow foxtail
Andropogon gerardi Vitman	Big bluestem

A. hallii Hach.	Sand bluestem
Bouteloua curtipendula (Michx.)Torr.	Sideoats gramma
B. gracilis (H.B.K.) Lag. ex Steud.	Blue gramma
Bromus inermis Leyss.	Smooth brome
B. marginatus Nees	Mountain brome
Bouchloe dactyloides (Nutt.) Engelm.	Buffalo grass
Calamovilfa longifolia (Hook.) Scribn.	Prairie sandreed grass
Dactylis glomerata L.	Orchardgrass
Deschampsia caespitosa (L.) Beauv.	Tufted hairgrass
Distichlis spicata (L.) Greene	Saltgrass
Elymus angustus Trin.[b]	Altai wildrye
E. canadensis L.	Canada wildrye
E. cinereus Scribn. & Merr.	Basin wildrye
E. gigantus Vahl	Mammoth wildrye
E. innovatus Beal	Hairy wildrye
E. junceus Fisch.[b]	Russian wildrye
E. salina Jones	Salina wildrye
E. triticoides Buckl.	Beardless wildrye
Festuca arundinacea Schreb.	Tall fescue
F. idahoensis Elmer	Idaho fescue
F. ovina L.	Sheep fescue
F. ovina var. *duriuscula* (L.) Koch	Hard sheep fescue
F. rubra L.	Red fescue
F. scabrella Torr.	Rough fescue
F. thruberi Vasey	Thurber fescue
Hilaria jamesii (Torr.) Benth.	Gallenta
Koeleria pyramidata (Lam.) Beauv.[c] (*K. cristata* (L.) Pers. of Hitchcock 1950)	Prairie junegrass
Lolium prenne L.	Perennial ryegrass
Oryzopsis hymenoides (Roem. & Schult.) Ricker	Indian ricegrass
Panicum virgatum L.	Switchgrass
Phalaris arundinacea L.	Reed canarygrass
Phleum alpinum L.	Alpine timothy
P. pratense L.	Timothy
Phragmites communis Trin.	Common reedgrass
Poa alpina L.	Alpine bluegrass
P. ampla Merr.	Big bluegrass
P. bulbosa L.	Bulbous bluegrass
P. canbyi (Scribn.) Piper	Canby bluegrass
P. compressa L.	Canada bluegrass
P. pratense L.	Kentucky bluegrass
P. sanbergii Vasey[b] (*P. secunda* Presl. of Hitchcock 1950)	Sandburg bluegrass
Puccinellia distans (L.) Parl.	Alkaligrass
Schizachyrium scoparium (Michx.) Nash[b] (*Andropogon scoparius* Michx. of Hitchcock 1950)	Little bluestem

Sitanion hystrix (Nutt.) J.G.Smith	Bottlebrush squirreltail
Sporobolus airoides (Torr.) Torr.	Alkali sacaton
S. cryptandrus (Torr.) Gray	Sand dropseed
Stipa columbiana Macoun.	Columbia needlegrass
S. comata Trin. & Rupr.	Needle-and-thread grass
S. lettermanii Vasey	Letterman needlegrass
S. viridula Trin.	Green needlegrass

Forbs

Scientific Name[d]	Common Name
Achillea millefolium L.	Western yarrow
Amorpha canescens Pursh	Lead plant
Artemisia ludoviciana Nutt.	Cudleaf sagewort
Astragalus ceramicus Sheld.	Painted milkvetch
A. cicer L.[e]	Cicer milkvetch
Coronilla varia L.[b]	Crown vetch
Dalea purpurea Vent.	Purple prairie clover
Echinacea pallida Nutt.[b]	Black samson
Glycyrrhiza lepidota Pursh	Wild licorice
Hedysarum boreale Nutt.	Northern sweetvetch
Helianthus annuus L.	Annual sunflower
H. maximiliani Schrad.	Maximillian sunflower
Heterotheca villosa (Pursh) Shinners	Hairy goldenaster
Kochia prostrata (L.) Schrad.[f]	Prostrate summercypress
Liatris punctata Hook.	Thickspike gayfeather
Linum lewisii Pursh	Lewis' flax
Lotus corniculatus L.	Birdsfoot trefoil
Lupinus sericeus Pursh	Silky lupine
Medicago lupulina L.	Black medic
M. sativa L.	Alfalfa
Melilotus alba Medic.	White sweetclover
M. officinalis (L.) Lam.	Yellow sweetclover
Onobrychis viciaefolia Scop.[b]	Sainfoin
Penstemon spp. Mitch.	Penstemon species
Ratibita columnifera (Nutt.) Woot. & Standl.	Prairie coneflower
Sanguisorba minor Scop.[e]	Small burnet
Solidago missouriensis Nutt.	Goldenrod
Sphaeralcea coccinea (Nutt.) Rydb.	Scarlet globemallow
Thermopsis rhombifolia (Nutt. ex Pursh) Nutt. ex Richards.	Golden pea
Trifolium hybridum L.	Alsike clover
T. pratense L.	Red clover
T. repens L.	White clover
Vicia americana Muhl. ex Willd.	American vetch

Shrubs	
Scientific Name[d]	Common Name
Amelanchier alnifolia (Nutt.) Nutt. ex M. Roem.	Serviceberry
Artemisia abrotanum L.	Oldman wormwood
A. cana Pursh	Silver sagebrush
A. dracunculus L.[b]	Green sagewort
A. frigida Willd.	Fringed sagewort
A. tridentata Nutt.	Big sagebrush
Atriplex canescens (Pursh) Nutt.	Fourwing saltbush
A. confertifolia (Torr. & Frem.) S.Wats	Shadscale
A. gardneri (Moq.) D. Dietr.	Gardner saltbush
Caragana arborescens Lam.[b]	Siberian peabush
Ceanothus velutinus Dougl. ex Hook.	Snowbrush ceanothus
Ceratoides lanata (Pursh) Howell	Winterfat
Cercocarpus lendifolius Nutt.	Curlleaf mountain mahogany
Chrysothamnus nauseosus (Pall. ex Pursh) Britt.	Rubber rabbitbush
C. viscidiflorus (Hook.) Nutt.	Green rabbitbush
Cornus stolonifera Michx.	Redosier dogwood
Cotoneaster acutifolius Turcz.[b]	Peking cotoneaster
Crataegus douglasii Lindl.	Black hawthorn
Grayia spinosa (Hook.) Moq.	Spiny hopsage
Juniperus communis L.	Common juniper
J. horizontalis Moench	Horizontal juniper
Potentilla fruticosa L.[b]	Shrubby cinquefoil
Prunus americana Marsh.	Wild plum
P. pumila L.	Sand cherry
P. virginiana L.	Common chokecherry
Purshia tridentata (Pursh) DC.	Antelope bitterbrush
Rhus trilobata Nutt.	Skunkbush sumac
Ribes aureum Pursh	Golden currant
R. cereum Dougl.	Wax currant
Rosa acicularis Lindl.[b]	Prickly rose
R. arkansana Porter	Prairie rose
R. woodsii Lindl.	Common wild rose
Sarcobatus vermiculatus (Hook.) Torr.	Black greasewood
Shepherdia argentea (Pursh) Nutt.[b]	Silver buffaloberry
Symphoricarpos occidentalis Hook.	Western snowberry
Yucca glauca Nutt.	Soapweed

Trees	
Scientific Name[d]	Common Name
Acer negundo L.	Box elder
Celtis occidentalis L.	Hackberry
Elaeagnus angustifolia L.	Russian olive
Fraxinus pennsylvanica Marsh.	Green ash
Pinus contorta Dougl. ex Loudon	Lodgepole pine

P. ponderosa Dougl. ex P. & C. Laws.	Ponderosa pine
Populus deltoides Marsh.	Eastern cottonwood
P. tremuloides Michx.	Quaking aspen
Pseudotsuga menziesii (Mirb.) Franco	Douglas fir
Quercus macrocarpa Michx.	Bur oak
Salix spp. L.	Willow
S. bebbiana Sarg.	Bebb willow
S. exigua Nutt.	Sandbar willow

[a] Nomenclature from Hitchcock 1950, except as noted.
[b] Nomenclature from Bailey and Bailey, 1976.
[c] Nomenclature from Yaeger et al.,1976.
[d] Nomenclature from Dorn, 1977.
[e] Nomenclature from Hitchcock and Cronquist, 1973.
[f] Nomenclature from SCS, 1988.

TABLE 35 Plant Species Used in Southwestern Revegetation Programs

Grasses	
Scientific Name[a]	Common Names
Agropyron cristatum (L.) Gaertn.	Crested wheatgrass
A. dasystachyum (Hook) Scribn.	Thickspike wheatgrass
A. desertorum (Fisch.) Schult.	Desert wheatgrass
A. elongatum (Host) Scribn.[b]	Tall wheatgrass
A. intermedium (Host) Beauv.[b]	Intermediate wheatgrass
A. riparium Scribn. & Smith	Streambank wheatgrass
A. sibiricum (Willd.) Beauv.[b]	Siberian wheatgrass
A. smithii Rydb.	Western wheatgrass
A. trachycaulum (Link) Malte	Slender wheatgrass
A. trichophorum (Link) Richt.[b]	Pubescent wheatgrass
Aristida purpurea Nutt.	Purple threeawn
Blepharoneuron tricholepis (Torr.) Nash	Pine dropseed
Bothriochloa barbinodis (Lag.) Herter[c]	Cane bluestem
B. ischaemum (L.) Keng[b]	Yellow bluestem
B. ischaemum var. *songaricus* (Rupr.) Celerier & Harlan[c]	King Ranch bluestem
Bouteloua curtipendula (Michx.) Torr.	Sideoats grama
B. eriopoda Torr.	Black grama
B. gacilis (H.B.K.) Steud.	Blue grama
Bromus marginatus Nees	Mountain brome
Buchloe dactyloides (Nutt.) Engelm.	Buffalograss
Cenchrus ciliaris L.[c]	Bufflegrass
Cynodon dactylon (L.) Pers.	Bermudagrass
Distichlis stricta (Torr.) Rydb.	Saltgrass
Elymus cinereus Scribn. & Merr.	Basin wildrye
E. junceus Fisch.	Russian wildrye
Eragrostis chloromelas Steud.	Boer lovegrass
E. curvula (Schrad.) Nees	Weeping lovegrass

E. intermedia Hitchc.	Plains lovegrass
E. lehmanniana Nees	Lehman's lovegrass
E. lehmanniana Nees X *E. trichophora* Coss. & Durieu[d]	Atherstone lovegrass
E. superba Peyritsch[e]	Wilman lovegrass
E. trichodes (Nutt.) Wood[b]	Sand lovegrass
E. trichophora Coss. & Durieu[e]	Cochise lovegrass
Festuca arizonica Vasey.	Arizona fescue
F. ovina L.	Sheep fescue
Hilaria jamesii (Torr.) Benth.	Galleta
Koeleria pyramidata (Lam.) Beauv.[c] (*K. cristata* (L.) Pers. of Kearney and Peebles 1951)	Prairie junegrass
Leptochloa dubia (H.B.K.) Nees.	Green sprangletop
Muhlenbergia montana (Nutt.) Hitchc.	Mountain muhly
M. porteri Scribn.	Bush muhly
M. wrightii Vasey[f]	Spike muhly
Oryzopsis hymenoides (Roem. & Schult.) Ricker	Indian ricegrass
Panicum antidotale Retz.	Blue panicgrass
P. coloratum L.	Kleingrass
P. obtusum H.B.K.	Vine mesquite
P. virgatum L.	Switchgrass
Poa pratensis L.	Kentucky bluegrass
P. sandbergii Vasey[b] (*P. secunda* Presl. of Kearney and Peebles 1951)	Sandburg bluegrass
Schizachyrium scoparius (Michx.) Nash[b]	Little bluestem
Schismus barbatus (L.) Thell.	Mediterranium grass
Setaria macrostachya H.B.K.	Plains bristlegrass
Sorghastrum nutans (L.) Nash.	Indian grass
Sorghum halepense (L.) Pers.	Johnsongrass
Sporobolus airoides Torr.	Alkali sacaton
S. cryptandrus (Torr.) Gray	Sand dropseed
S. flexuosus (Thurb.) Rydb.	Mesa dropseed
S. giganteus Nash	Giant dropseed
Stipa comata Trin. & Rupr.	Needle-and-thread grass
S. viridula Trin.	Green needlegrass
Trichachne californica (Benth.) Chase	Arizona cotton-top

Forbs

Scientific Name[a]	Common Names
Achillea lanulosa Nutt. (indistinguishable from *A. millefollium* L.)	Western yarrow
Artemisia ludoviciana Nutt.	Cudleaf sagewort
Aster spp. L.[b]	Aster
Astragalus cicer L.[h]	Cicer milkvetch
Erodium cicutarium (L.) L'Her.	Filaree
Eriogonum alatum Torr.	Wild buckwheat
Linum lewisii Pursh	Lewis' flax

Medicago sativa L.	Alfalfa
Melilotus alba Medic.	White sweetclover
M. officinalis (L) Lam.	Yellow sweetclover
Oenothera hookerii Torr. & Gray	Hookers evening primrose
Penstemon spp. Mitch.[b]	Beardtongue
P. strictus Benth.	Rocky mountain penstemon
Petalostemon purpureum (Vent.) Rybd.	Purple prairie clover
Ratibida columnaris (Sims) D. Don.	Mexican hat
Sanguisorba minor Scop.	Small burnet
Sphaeralcea grossulariaefolia (Hook. & Arn.) Rydb.	Scarlet globemallow

Shrubs

Scientific Name[a]	Common Names
Acacia constrica Benth.	Whitethorn
A. greggii Gray	Catclaw
Amelanchier utahensis Koehne.	Utah serviceberry
Artemisia frigida Willd.	Fringed sagewort
Atriplex canescens (Pursh) Nutt.	Fourwing saltbush
A. confertifolia (Torr. & Frem.) S.Wats	Shadscale
A. corrugata Wats.	Mat saltbush
A. lentiformis (Torr.) Wats.	Quailbush
A. obovata Moq.	Broadscale
A. polycarpa (Torr.) Wats.	Desert saltbush
A. semibaccata R. Br.	Australian saltbush
Calliandra eriophylla Benth.	False mesquite
Ceratoides lanata (Pursh) Howell[i] (*Eurotia lanata* (Pursh) Moq. of Kearney and Peebles 1951)	Winterfat
Cercidium floridum Benth	Blue Palo verde
C. microphyllum (Torr.) Rose and Johnson	Little leaf Palo verde
Cercocarpus montanus Raf.	Mountain mahogany
Chrysothamnus nauseosus (Pall.) Britt.	Rubber rabbitbrush
Cowania mexicana var. *stansburiana* (Torr.) Jepson	Stansbury cliffrose
Eriogonum wrightii Torr.	Shrubby buckwheat
Ephedra spp. L.	Mormontea
Forestiera neomexicana Gray	New Mexico olive
F. pubescens Nutt.[h]	Prairie adelia
Franseria deltoidea Torr.	Triangle leaf bursage
Larrea tridentata (DC.) Coville	Creosote bush
Menodora scabra Gray	Rough menodora
Parkinsonia aculeata L.	Mexican Palo verde
Prosopis alba Griseb.[j]	Argentine mesquite
P. chilensis (Mol.) Stuntz.[b]	Chilean mesquite
P. juliflora (Swartz) DC.	Common mesquite
P. velutina Woot.[k] (*P. juliflora* var. *velutina* (Woot.) Sarg. of Kearney and Peebles 1951)	Velvet mesquite

Purshia tridentata (Pursh) DC	Antelope bitterbrush
Rhus trilobata Nutt.	Skunkbush sumac
Ribes cereum Dougl.	Wax currant
Rosa woodsii Lindl.[b]	Common wild rose
Sarcobatus vermiculatus (Hook.) Torr.	Black greasewood
Simmondsia chinensis (Link) Schneid.	Jojoba
Symphoricarpos occidentalis Hook.[b]	Western snowberry
Tamarix aphylla (L.) Karst.	Evergreen tamarix
T. pentandra Pall.	Saltcedar

Trees

Scientific Name[a]	Common Names
Ailanthus altissima (Mill.) Swingle	Tree-of-heaven
Eucalyptus camaldulensis Dehnh.[b]	Red gum
E. microtheca F. J.Muell[b]	Coolibah
Juniperus osteosperma (Torr.) Little	Utah juniper
Pinus cembroides Zucc.	Nut pine
P. edulis Engelm.	Pinyon pine
P. monophylla Torr. & Frem.	Single leaf pinyon
P. ponderosa Dougl. Lawson	Ponderosa pine
Populus fremontii Wats.	Fremont cottonwood
Quercus gambelii Nutt.	Gamble oak
Robinia neomexicana Gray	New Mexico locust

[a] Nomenclature from Kearney and Peebles, 1951 except as noted.
[b] Nomenclature from Bailey and Bailey, 1976.
[c] Nomenclature from Gould, 1975.
[d] Nomenclature from S.C.S., 1983.
[e] Nomenclature from S.C.S., 1982.
[f] Nomenclature from Hitchcock, 1950.
[g] Nomenclature from Holt et al., no date.
[h] Nomenclature from Hitchcock and Cronquist, 1973.
[i] Nomenclature from Dorn, 1977.
[j] Nomenclature from Hook and Jackson, 1895.
[k] Nomenclature from Martin and Hutchins, 1980.

REFERENCES

1. Bailey, L. H. and Bailey, E. Z., *Hortus Third*, Macmillan, New York, 1976.
2. Dorn, R. D., *Manual of the Vascular Plants of Wyoming*, Vol. I and II, Garland Publishing, New York, 1977.
3. Gould, F. W., *The Grasses of Texas*, Texas A & M University Press, College Station, 1975.
4. Hitchcock, A. S., Manual of the Grasses of the United States, 2nd ed., U.S. Government Printing Office, Washington, D.C., 1950.

5. Hitchcock, C. L. and Cronquist, A., *Flora of the Pacific Northwest*, University of Washington Press, Seattle, 1973.
6. Holt, E. C., Novosad, A. C., Pratt, J. N., and Davis, A. G., *Kleingrass*, Agric. Exp. Sta., Texas A & M University, College Station.
7. Hooke, J. P. and Jackson, B. D., *Index Kewensis*, Tomus II, D-Z, Oxford University Press, New York, 1895.
8. Kearney, T. H. and Peebles, R. H., *Arizona Flora*, University of California Press, Berkeley, 1951.
9. Lawrence, T., Alberta Wild Ryegrass, Publ. 1602, Inf. Serv., Agriculture Canada, Ottawa, 1981.
10. Martin, W. C. and Hutchins, C. R., *A Flora of New Mexico*, Verlag, Germany, 1980.
11. Mohlenbrock, R. H., *Guide to the Vascular Flora of Illinois*, Southern Illinois University Press, Carbondale, 1986.
12. Rehder, A., *Manual of Cultivated Trees and Shrubs*, 2nd ed., Macmillan, New York, 1954.
13. Soil Conservation Service, Cochise Atherstone Lovegrass, S.C.S. Prog. Aid No. 1319, U.S. Department of Agriculture, Tucson, AZ, 1983.
14. Soil Conservation Service, National List of Scientific Plant Names, Vol. 1, SCS-TP-159, U.S. Department of Agriculture, Washington, D.C., 1982.
15. Soil Conservation Service, Plant Materials Handbook, SCS Plant Materials Center, U.S. Department of Agriculture, Denver, CO, 1988.
16. Yaeger, R. R., Leininger, W. C., and Ryerson, D. E., *Common Montana Range Plants*, 2nd ed., Montana State University, Bozeman, 1976.

INDEX

A

Abandoned mine lands (AML), applied revegetation techniques and, 204–207
AB-DTPA; see Ammonium bicarbonate-DTPA
Acacia constricta, in Southwestern revegetation programs, 225
Acacia greggii, in Southwestern revegetation programs, 225
Access roads, applied revegetation techniques and, 176–178
Acer negundo, in Northern Great Plains revegetation programs, 222
Acer rubrum
  in Eastern revegetation programs, 215
  in Midwestern revegetation programs, 218
Acer saccharinum
  in Eastern revegetation programs, 215
  in Midwestern revegetation programs, 218
Acer saccharum
  in Eastern revegetation programs, 215
  in Midwestern revegetation programs, 218
Achillea lanulosa, in Southwestern revegetation programs, 224
Achillea millefolium
  minesoils and, 165
  in Northern Great Plains revegetation programs, 221
  in Southwestern revegetation programs, 224
Acid-base accounts
  lime and, 98
  tailings and, 181
Acid-base analysis, of overburden, 24
Acid extraction, of manganese, 35
Acid-generating capacity, estimating, 98
Acid mine drainage, sulfur in, 24
Acid soils
  boron in, 30
  cadmium in, 31
  calcium in, 21–22
  lime in, 96–97
  nickel in, 39
  oxidation and, 17
  pH and, 5–6
  phosphorus in, 82–83
  sulfur in, 24
  tailings and, 178–181, 183–188
Acid spoil, topsoiling and, 69
Acid wastes, manure and, 94
Aeration, soil, texture influence on, 10
Aerial broadcasting, in surface manipulation, 74
Aerial seeding, characterization of, 124
Agricultural lands, maximum trace element loadings for, 97
Agropyron cristatum
  in Northern Great Plains revegetation programs, 219
  seeding and planting of, 110
  in Southwestern revegetation programs, 223
Agropyron dasystachyum
  in Northern Great Plains revegetation programs, 115, 219
  in Southwestern revegetation programs, 223
Agropyron desertorum
  in Northern Great Plains revegetation programs, 219
  in Southwestern revegetation programs, 223
Agropyron elongatum
  in Northern Great Plains revegetation programs, 219
  in Southwestern revegetation programs, 223
Agropyron inerme, in Northern Great Plains revegetation programs, 219
Agropyron intermedium
  in Northern Great Plains revegetation programs, 219
  in Southwestern revegetation programs, 223
Agropyron riparium
  in Northern Great Plains revegetation programs, 219
  in sand and gravel mines, 197
  in Southwestern revegetation programs, 223
Agropyron sibiricum
  in Northern Great Plains revegetation programs, 219

in Southwestern revegetation programs, 223
Agropyron smithii
in Northern Great Plains revegetation programs, 115, 219
in Southwestern revegetation programs, 223
Agropyron spicatum
minesoils and, 165
in Northern Great Plains revegetation programs, 219
Agropyron trachycaulum
in Northern Great Plains revegetation programs, 219
in Southwestern revegetation programs, 223
waste rock dumps and, 176
Agropyron trichophorum
in Northern Great Plains revegetation programs, 219
in Southwestern revegetation programs, 223
tailings and, 187
Agrostis alba
in Eastern revegetation programs, 213
in Midwestern revegetation programs, 217
in Northern Great Plains revegetation programs, 219
tailings and, 180
Agrostis scabra, in Northern Great Plains revegetation programs, 219
Agrostis spp., aluminum in, 26
Agrostis stolonifera, in Northern Great Plains revegetation programs, 219
Agrostis tenuis
in Eastern revegetation programs, 213
in Northern Great Plains revegetation programs, 219
Ailanthus altissima
in Midwestern revegetation programs, 218
in Southwestern revegetation programs, 226
Air pollution; see also Pollution
arsenic and, 29
cadmium and, 31
mercury and, 36
sulfur toxicity and, 23
Alfalfa
in Eastern revegetation programs, 214
in Midwestern revegetation programs, 218
minesoils and, 167
nitrogen in, 18
in Northern Great Plains revegetation programs, 221
phosphorus in, 20
in Southwestern revegetation programs, 225
Algae, topsoiling and, 67
Alkaline soils
copper in, 33
nickel in, 39
pH of, 6–8
sodium in, 7
tailings and, 178–179, 185
Alkali sacaton
in Northern Great Plains revegetation programs, 221
in Southwestern revegetation programs, 224
Alluvial deposits, topsoiling and, 66
Aluminum
calcium and, 21
coal ash and, 201
deficiency of, 25
tailings and, 180
topsoiling and, 68
toxicity of, 25–27
Alnus glutinosa
in Eastern revegetation programs, 215
in Midwestern revegetation programs, 218
Alopecurus arundinaceus, in Northern Great Plains revegetation programs, 219
Alopecurus pratensis, in Northern Great Plains revegetation programs, 219
Alpha diversity, definition of, 111
Alsike clover
in Eastern revegetation programs, 214
in Midwestern revegetation programs, 218
in Northern Great Plains revegetation programs, 221
Amelanchier alnifolia, in Northern Great Plains revegetation programs, 222
Amelanchier utahensis, in Southwestern revegetation programs, 225
Amendments
site preparation and, 76, 79–100
tailings and, 183–184
American elder, in Eastern revegetation programs, 215
American plum, in Eastern revegetation programs, 215
American sycamore
in Eastern revegetation programs, 216
in Midwestern revegetation programs, 219

American vetch
in Northern Great Plains revegetation programs, 221
waste rock dumps and, 175
Amino acids, sulfur in, 23
AML; see Abandoned mine lands
Ammonium acetate extraction
of arsenic, 28
of lead, 35
of manganese, 35
of molybdenum, 37
of potassium, 21
of sodium, 43
of trace elements, 26
of zinc, 16
Ammonium bicarbonate-DTPA (AB-DTPA) extraction
of boron, 31
of cadmium, 30
of copper, 32
of lead, 35
of manganese, 35
of phosphorus, 20
of selenium, 40–41
Ammonium carbonate extraction, of molybdenum, 37
Ammonium nitrate, in fertilizers, 80
Ammonium oxalate-oxalic acic extraction, of molybdenum, 37
Ammonium phosphate, in fertilizers, 80
Ammonium sulfate, in fertilizers, 80
Amorpha canescens
in Midwestern revegetation programs, 217
in Northern Great Plains revegetation programs, 221
Amorpha fruticosa, in Eastern revegetation programs, 215
Amur honeysuckle, in Eastern revegetation programs, 215
Amur privet, in Eastern revegetation programs, 215
Andropogon gerardi
in Eastern revegetation programs, 213
in Midwestern revegetation programs, 217
in Northern Great Plains revegetation programs, 219
Andropogon hallii, in Northern Great Plains revegetation programs, 220
Andropogon scoparius
in Eastern revegetation programs, 214
in Midwestern revegetation programs, 217
in Northern Great Plains revegetation programs, 220
Angle of repose, in slope reconstruction, 61, 63
Anhydrous ammonia, in fertilizers, 80
Anions, cation exchange capacity and, 13
Annuals
plant succession and, 2
temporary stabilization and, 126
Annual sunflower
in Eastern revegetation programs, 214
in Midwestern revegetation programs, 217
in Northern Great Plains revegetation programs, 221
Antelope bitterbrush
in Northern Great Plains revegetation programs, 222
in Southwestern revegetation programs, 226
AOC; see Approximate original contour
Applied revegetation techniques; see specific techniques
Approximate original contour (AOC)
impoundments and, 152, 156
inplace spoil ridges and, 156
recontouring and, 59
Aquatic vegetation, impoundments and, 156
Argentine mesquite, in Southwestern revegetation programs, 225
Aristida purpurea, in Southwestern revegetation programs, 223
Arizona, tailings in, 178
Arizona cotton-top, in Southwestern revegetation programs, 224
Arizona fescue, in Southwestern revegetation programs, 223
Arrhenatherum elatius, in Eastern revegetation programs, 213
Arsenic
coal ash and, 201
toxicity of, 27–29
Artemisia abrotanum, in Northern Great Plains revegetation programs, 222
Artemisia cana
minesoils and, 165
in Northern Great Plains revegetation programs, 222
Artemisia dracunculus, in Northern Great Plains revegetation programs, 222
Artemisia frigida
minesoils and, 165
in Northern Great Plains revegetation programs, 222
in Southwestern revegetation programs, 225

Artemisia ludoviciana
in Northern Great Plains revegetation programs, 221
in Southwestern revegetation programs, 224
Artemisia spp., community diversity and, 145
Artemisia tridentata
minesoils and, 165
in Northern Great Plains revegetation programs, 222
waste rock dumps and, 176
Arthropods, topsoiling and, 67
Ash disposal ponds, applied revegetation techniques and, 201–204
Asia, composting in, 95
Aspect
applied revegetation techniques and, 165
metal mine wastes and, 169
plant species selection and, 114–115
Aster spp.
selenium in, 42
in Southwestern revegetation programs, 224
Astragalus ceramicus, in Northern Great Plains revegetation programs, 221
Astragalus cicer
in Eastern revegetation programs, 214
in Midwestern revegetation programs, 217
minesoils and, 167
in Northern Great Plains revegetation programs, 221
in Southwestern revegetation programs, 224
Astragalus spp., selenium in, 41
Atherstone lovegrass, in Southwestern revegetation programs, 223
Atriplex canescens
boron in, 29
in Northern Great Plains revegetation programs, 222
in Southwestern revegetation programs, 225
Atriplex confertifolia
in Northern Great Plains revegetation programs, 222
in Southwestern revegetation programs, 225
Atriplex corrugata, in Southwestern revegetation programs, 225
Atriplex gardneri, in Northern Great Plains revegetation programs, 222
Atriplex lentiformis, in Southwestern revegetation programs, 225
Atriplex nummularia, sodium in, 43
Atriplex obovata, in Southwestern revegetation programs, 225
Atriplex polycarpa, in Southwestern revegetation programs, 225
Atriplex semibaccata, in Southwestern revegetation programs, 225
Atriplex spp., selenium in, 41
Atriplex vesicaria, sodium in, 43
Australia
gouging in, 73
tailings in, 178
Australian saltbush, in Southwestern revegetation programs, 225
Austrian pine
in Eastern revegetation programs, 216
in Midwestern revegetation programs, 218
Autumn olive
in Eastern revegetation programs, 215
in Midwestern revegetation programs, 218
Auxin stability, zinc in, 43
Avena sativa, in mulch, 90

B

Backfilling
impoundments and, 156
metal mine wastes and, 169
Bacteria
mercury compounds and, 36
nitrogen fertilizers and, 81–82
organic matter and, 9
topsoiling and, 67–68
Bahiagrass, in Eastern revegetation programs, 213
Bald cypress
in Eastern revegetation programs, 216
in Midwestern revegetation programs, 219
Bareroot stock, transplantation of, 128–129, 131–132, 136–137
Bark fragments, as mulch, 89–90
Barnyard grass
in Eastern revegetation programs, 213
in Midwestern revegetation programs, 217
Basin wildrye, in Southwestern revegetation programs, 223
Basswood, in Eastern revegetation programs 216
B concentration, of soil, 13
Beardless wheatgrass, in Northern Great Plains revegetation programs, 219

Beardtongue
in Eastern revegetation programs, 214
in Southwestern revegetation programs, 225
Bebb willow, in Northern Great Plains revegetation programs, 223
Bedrock, topsoiling and, 67
Bentgrass species, aluminum in, 26
Bentonite mining
applied revegetation techniques and, 191–195
land rehabilitation and, 90
Bermudagrass
in Eastern revegetation programs, 213
in Midwestern revegetation programs, 217
in Southwestern revegetation programs, 223
Beta diversity, definition of, 111
Betula lenta, in Eastern revegetation programs, 215
Betula nigra
in Eastern revegetation programs, 215
in Midwestern revegetation programs, 218
Betula papyrifera, in Eastern revegetation programs, 215
Betula pendula, in Eastern revegetation programs, 215
Betula populifolia, in Eastern revegetation programs, 215
Bicarbonate extraction
of arsenic, 28
of phosphorus, 20
Biennials
plant succession and, 2
temporary stabilization and, 126
Big bluestem
in Eastern revegetation programs, 213
in Midwestern revegetation programs, 217
Big sagebrush
minesoils and, 165–166
in Northern Great Plains revegetation programs, 222
waste rock dumps and, 174–176
Big Sky Mine, Montana, impoundments and, 152
Biomass, above ground, green manure and, 95
Birdsfoot trefoil
in Eastern revegetation programs, 214
in Midwestern revegetation programs, 217
in Northern Great Plains revegetation programs, 221
tailings and, 187
Black cherry
in Eastern revegetation programs, 216
in Midwestern revegetation programs, 219
Black-eyed cowpea, in Eastern revegetation programs, 214
Black grama, in Southwestern revegetation programs, 223
Black greasewood
in Northern Great Plains revegetation programs, 222
in Southwestern revegetation programs, 226
Black hawthorn, in Northern Great Plains revegetation programs, 222
Black Hills, South Dakota, bentonite mining in, 191–194
Black locust
in Eastern revegetation programs, 216
in Midwestern revegetation programs, 219
Black medic, in Northern Great Plains revegetation programs, 221
Black samson, in Northern Great Plains revegetation programs, 221
Black Thunder Mine, Wyoming, rock ledges and, 158
Black walnut
in Eastern revegetation programs, 216
in Midwestern revegetation programs, 218
Black willow, in Eastern revegetation programs, 216
Blepharoneuron tricholepis, in Southwestern revegetation programs, 223
BLM; see Bureau of Land Management
Blueberry, in Eastern revegetation programs, 215
Bluebunch wheatgrass
highway shoulders and, 200
minesoils and, 165–166
in Northern Great Plains revegetation programs, 219
waste rock dumps and, 174–176
Blue flax
minesoils and, 166–167
waste rock dumps and, 175–176
Blue grama
in Eastern revegetation programs, 213
highway shoulders and, 200
minesoils and, 166
in Southwestern revegetation programs, 223

Blue Palo verde, in Southwestern revegetation programs, 225
Blue panicgrass, in Southwestern revegetation programs, 224
Blue sage, in Eastern revegetation programs, 214
Bluff extensions, landscape diversity and, 151–152
Boer lovegrass, in Southwestern revegetation programs, 223
Boron
  applied revegetation techniques and, 164
  ash disposal ponds and, 202, 204
  calcium and, 21
  coal ash and, 201
  deficiency of, 25, 29–30
  soil extraction of, 16–17
  tailings and, 181
  topsoiling and, 68
  toxicity of, 21, 25, 29–30
Bothriochloa barbinodis, in Southwestern revegetation programs, 223
Bothriochloa caucasia, in Eastern revegetation programs, 213
Bothriochloa ischaemum, in Southwestern revegetation programs, 223
Bothriochloa ischaemum var. *songaricus*, in Southwestern revegetation programs, 223
Bouteloua curtipendula
  in Eastern revegetation programs, 213
  minesoils and, 166
  in Northern Great Plains revegetation programs, 220
  in Southwestern revegetation programs, 223
Bouteloua eripoda, in Southwestern revegetation programs, 223
Bouteloua gracilis
  in Eastern revegetation programs, 213
  minesoils and, 166
  in Northern Great Plains revegetation programs, 220
  in Southwestern revegetation programs, 223
Box elder, in Northern Great Plains revegetation programs, 222
Bray extraction
  of arsenic, 28
  of phosphorus, 20
Breaks, inplace spoil ridges and, 157
Brillion broadcast seeder, representation of, 123
Bristly locust
  in Eastern revegetation programs, 215
  in Midwestern revegetation programs, 218
Broadcast seeder, representation of, 122–123
Broadcast seeding
  characterization of, 122–124
  seedbed preparation and, 112
  seed rates and, 118
surface manipulation and, 64, 72–74, 122–124
Broadscale, in Southwestern revegetation programs, 225
Broadleaf plants
  calcium in, 21
  copper in, 33
  magnesium in, 22
Brome grasses, in native hay mulches, 92
Bromus arvensis, in Eastern revegetation programs, 213
Bromus inermis
  in Eastern revegetation programs, 213
  manganese in, 35
  in Midwestern revegetation programs, 217
  in Northern Great Plains revegetation programs, 220
Bromus japonicus, in native hay mulches, 92
Bromus marginatus
  in Northern Great Plains revegetation programs, 220
  in Southwestern revegetation programs, 223
  waste rock dumps and, 175
Bromus tectorum, in native hay mulches, 92
Buchloe dactyloides
  in Eastern revegetation programs, 213
  in Northern Great Plains revegetation programs, 220
  in Southwestern revegetation programs, 223
Buckwheat, in Eastern revegetation programs, 214
Buffalograss
  in Eastern revegetation programs, 213
  in Southwestern revegetation programs, 223
Buffering capacity, of soil, 13
Bufflegrass, in Southwestern revegetation programs, 223
Bulk density, of soil, 10–12
Bulldozer attachment, for creating catchment basins on minesoil, 75
Bureau of Land Management (BLM), metal mining regulation by, 168–169

Burlap, in erosion control blankets, 92, 94
Bur oak
in Eastern revegetation programs, 216
in Midwestern revegetation programs, 219
in Northern Great Plains revegetation programs, 223
Bush muhly, in Southwestern revegetation programs, 224

C

Ca^{2+}, exchange capacity of, 12
Cadmium
deficiency of, 25
federal recommendations for maximum loading of in agricultural lands, 97
nitrate reduction and, 19
sewage sludge and, 96
tailings and, 180
topsoiling and, 68
toxicity of, 25, 30–31
Calamovilfa longifolia
in Northern Great Plains revegetation programs, 220
in sand and gravel mines, 197
Calcium
aluminum and, 26–27
amendments and, 97–99
boron and, 30
coal ash and, 201
deficiency of, 22–23
hydrogen ions and, 26–27
iron and, 34
magnesium and, 21–22
sodium and, 7, 43
tailings and, 180–181, 186
Calcium carbonate; see Lime
Calcium chloride
bentonite mining and, 194–195
sodicity and, 99
Calcium oxide; see Lime
California, grass tetany in, 23
Calliandra eriophylla, in Southwestern revegetation programs, 225
Canada
acid-base balance in soils of, 24
nickel in, 39
soil pH of mine sites in, 6
tailings in, 178
Canada bluegrass
in Eastern revegetation programs, 214
highway shoulders and, 200
in Midwestern revegetation programs, 217
tailings and, 187
Cane bluestem, in Southwestern revegetation programs, 223
Caragana arborescens
in Eastern revegetation programs, 215
in Midwestern revegetation programs, 218
in Northern Great Plains revegetation programs, 222
Carbon
amendments and, 88–90
compost and, 95–96
manure and, 95
nitrogen and, 9
soil organic matter and, 9–10
Carex filifolia, minesoils and, 165
Carriers, in drill seeding, 121
Carya illinoensis
in Eastern revegetation programs, 215
in Midwestern revegetation programs, 218
Carya spp.
in Eastern revegetation programs, 215
in Midwestern revegetation programs, 218
Cassia fasiculata, in Eastern revegetation programs, 214
Castanea mollissima, in Eastern revegetation programs, 215
Catalpa speciosa, in Eastern revegetation programs, 215
Catalysis, manganese in, 35
Catclaw, in Southwestern revegetation programs, 225
Cation exchange capacity (CEC)
access roads and, 176
bentonite and, 191
compost and, 95
organic matter and, 8
plant establishment and, 12–13
of sand and gravel pits, 195–196
sewage sludge and, 97
sodium hazard estimation and, 7
texture influence on, 10
topsoiling and, 66
waste rock and, 170
Caucasian bluestem, in Eastern revegetation programs, 213
Ceanothus velutinus
in Northern Great Plains revegetation programs, 222
waste rock dumps and, 176
CEC; see Cation exchange capacity
Cell wall structure, calcium role in, 21

Celtis occidentalis
in Eastern revegetation programs, 216
in Midwestern revegetation programs, 218
in Northern Great Plains revegetation programs, 222
Cenchrus ciliaris, in Southwestern revegetation programs, 223
Ceratoides lanata
minesoils and, 167
in Northern Great Plains revegetation programs, 222
in Southwestern revegetation programs, 225
Cercidium floridum, in Southwestern revegetation programs, 225
Cercidium microphyllum, in Southwestern revegetation programs, 225
Cercocarpus lendifolius, in Northern Great Plains revegetation programs, 222
Cercocarpus montanus, in Southwestern revegetation programs, 225
Cereal grains, as nurse crops, 126
Cereal grain stems, in mulch, 90
Certification programs, seed rates and, 117
Chestnut oak, in Midwestern revegetation programs, 219
Chile, tailings in, 178
Chilean mesquite, in Southwestern revegetation programs, 225
Chinese chestnut, in Eastern revegetation programs, 216
Chiselling
surface manipulation and, 70–72, 75–76, 83
tillage and, 112
Chlorine
ionic forms of, 16
toxicity of, 31–32
Chlorophyll
iron and, 33
magnesium and, 22
Chloroplast development, zinc in, 43
Chokecherry, in Eastern revegetation programs, 215
Chromium, sewage sludge and, 96
Chrysothamnus nauseosus
minesoils and, 167
in Northern Great Plains revegetation programs, 222
in Southwestern revegetation programs, 225
Chrysothamnus viscidflorus, in Northern Great Plains revegetation programs, 222
Cicer milkvetch
in Eastern revegetation programs, 214
in Midwestern revegetation programs, 217
minesoils and, 166–167
in Northern Great Plains revegetation programs, 221
in Southwestern revegetation programs, 224
Cinnabar vein, plants in soil over, 37
Clay cap
ash disposal ponds and, 203
tailings and, 184–187
Clay soils
diameter range of, 10
plant species selection and, 114
porosity of, 12
sodium concentrations in, 8
topsoiling and, 69
Climax community, in plant succession, 2
Coal ash, chemical constituents in, 201
Coal beds, selenium in, 40
Coal fields
acid-base balance in, 24
arsenic in, 29
bluff extensions and, 151–152
lead in, 35
lime in, 98
nitrogen fertilizers and, 81
phosphorus fertilizers and, 84
selenium in, 42
Coal mines
applied revegetation techniques and, 161–167
disking and, 71
phosphorus fertilizers and, 83
plant species selection and, 116
recontouring and, 59
rippers and, 73
rock ledges and, 157–158
slope reconstruction and, 63
soil pH in, 6
Coal minespoils, abandoned, 206
Cochise lovegrass, in Southwestern revegetation programs, 223
Coconut fiber, in erosion control blankets, 92–94
Colonial bentgrass
in Eastern revegetation programs, 213
in Northern Great Plains revegetation programs, 219

Colorado
- copper in, 32
- manganese in, 35
- nitrogen fertilizers and, 81
- slope reconstruction in, 61
- zinc in, 43

Colorimetric determination
- of nitrates, 19
- of phosphorus, 20

Columbia needlegrass
- in Northern Great Plains revegetation programs, 221
- waste rock dumps and, 175

Common chokecherry, in Northern Great Plains revegetation programs, 222

Common juniper, in Northern Great Plains revegetation programs, 222

Common lespedeza
- in Eastern revegetation programs, 214
- in Midwestern revegetation programs, 217

Common lupine, waste rock dumps and, 174–176

Common mesquite, in Southwestern revegetation programs, 225

Common vetch, in Eastern revegetation programs, 214

Common wild rose
- in Northern Great Plains revegetation programs, 222
- in Southwestern revegetation programs, 226

Community diversity, enhancement of, 144–148

Compaction, access roads and, 177

Companion crops, in temporary stabilization, 125–126

Competition
- planting success and, 118, 128, 137
- vegetation diversity and, 141

Compost, as organic amendment, 95–96, 183

Concave slopes, slope reconstruction and, 62, 64

Condensation traps, against water deficit-induced transplant loss, 133–134

Containerized stock, transplantation of, 128–132

Contour mining, applied revegetation techniques and, 161

Convex slopes, slope reconstruction and, 62

Coolibah, in Southwestern revegetation programs, 226

Copper
- applied revegetation techniques and, 164
- deficiency of, 25, 33
- federal recommendations for maximum loading of in agricultural lands, 97
- ionic forms of, 16
- molybdenum and, 33, 37–38
- sewage sludge and, 96
- tailings and, 180
- topsoiling and, 68
- toxicity of, 25, 32

Cornus ammomum
- in Eastern revegetation programs, 215
- in Midwestern revegetation programs, 218

Cornus florida
- in Eastern revegetation programs, 216
- in Midwestern revegetation programs, 218

Cornus racemosa
- in Eastern revegetation programs, 215
- in Midwestern revegetation programs, 218

Cornus stolonifera
- in Eastern revegetation programs, 215
- in Midwestern revegetation programs, 218
- in Northern Great Plains revegetation programs, 222

Coronilla varia
- in Eastern revegetation programs, 214
- in Midwestern revegetation programs, 217
- in Northern Great Plains revegetation programs, 221

Cotoneaster acutifolius, in Northern Great Plains revegetation programs, 222

Coversoil
- seedbed preparation and, 111
- topsoiling and, 66–67, 69

Cowania mexicana var. *stansburiana*, in Southwestern revegetation programs, 225

Cow manure, C:N ratio for, 89

Crab apple, in Eastern revegetation programs, 216

Crataegus douglassi, in Northern Great Plains revegetation programs, 222

Crataegus spp.
- in Eastern revegetation programs, 215
- in Midwestern revegetation programs, 218

Creeping bentgrass, in Northern Great Plains revegetation programs, 219

Creeping foxtail, in Northern Great Plains revegetation programs, 219

Creosote bush, in Southwestern revegetation programs, 225

Crested wheatgrass
highway shoulders and, 200
minesoils and, 164
in Northern Great Plains revegetation programs, 219
seeding and planting of, 110
in Southwestern revegetation programs, 223
Crimping, in securing mulches, 85–87, 91–92
Crimson clover, in Eastern revegetation programs, 214
Critana thickspike wheatgrass, in northern Great Plains revegetation programs, 115
Crown vetch
in Eastern revegetation programs, 214
in Midwestern revegetation programs, 217
in Northern Great Plains revegetation programs, 221
Cudleaf sagewort
in Northern Great Plains revegetation programs, 221
in Southwestern revegetation programs, 224
Cultipacker, behind disk, 120
Curlleaf mountain mahogany, in Northern Great Plains revegetation programs, 222
Cuttings, in establishing shrubs or trees, 133
Cynodon dactylon
in Eastern revegetation programs, 213
in Midwestern revegetation programs, 217
in Southwestern revegetation programs, 223

D

Dactylis glomerata
in Eastern revegetation programs, 213
in Midwestern revegetation programs, 217
in Northern Great Plains revegetation programs, 220
phosphate mines and, 191
Dalea purpurea
in Eastern revegetation programs, 214
minesoils and, 167
in Northern Great Plains revegetation programs, 221
Dallis grass, in Eastern revegetation programs, 213
Dark, germination and, 118
Deertongue
in Eastern revegetation programs, 213
in Midwestern revegetation programs, 217
Denitrification, nitrate ions and, 18
Deposition, drainage basin integrity and, 64
Deschampsia caespitosa
in Eastern revegetation programs, 213
in Midwestern revegetation programs, 217
in Northern Great Plains revegetation programs, 115, 220
Desert saltbush, in Southwestern revegetation programs, 225
Desert wheatgrass
in Northern Great Plains revegetation programs, 219
in Southwestern revegetation programs, 223
Desmanthus illinoensis, in Eastern revegetation programs, 214
Dessication, planting failure and, 136
Development, soil, plant species selection and, 116
Dewatering, tailings and, 184
Diammonium phosphate, in fertilizers, 80
Dieoffs, planting failure and, 137
Diethylenetriaminepentaacetic acid (DTPA) extraction
of boron, 31
of cadmium, 30
of copper, 32
of iron, 34
of lead, 35
of manganese, 35
of nickel, 39
of sodium, 43
of zinc, 16, 43
Diffusion barriers
ash disposal ponds and, 203
tailings and, 185
Direct-haul topsoil
community diversity and, 145–147
nutrient cycles of, 68
Disking
surface manipulation and, 70–72, 83
tillage and, 112
Distichlis spicata
chlorine in, 31
in Northern Great Plains revegetation programs, 220
Distichlis stricta, in Southwestern revegetation programs, 223

Diversity
applied revegetation techniques and, 165
definition of, 110–111
enhancement of, 141–158
plant species selection and, 116
Dormant fall seedings, characterization of, 127
Douglas fir
in Eastern revegetation programs, 216
in Northern Great Plains revegetation programs, 223
Dozer basins, digging of, in surface manipulation, 70, 73–77
Drainage basins, integrity of, recontouring and, 64–66
Drill pads, applied revegetation techniques and, 204
Drill seeder, representation of, 119
Drill seeding
characterization of, 119–121
seedbed preparation and, 112
seed rates and, 118
surface manipulation and, 72, 74, 92, 119–121
Drinking water, lead in, 34
Drought
applied revegetation techniques and, 163
planting failure and, 136
topsoiling and, 69
DTPA; see Diethylenetriaminepentaacetic acid
Dyes, in hydromulching, 87
Dynamic equilibrium, topsoiling and, 69

E

Earth moving
drainage basin integrity and, 64
slope reconstruction and, 62
Eastern cottonwood
in Eastern revegetation programs, 216
in Midwestern revegetation programs, 219
in Northern Great Plains revegetation programs, 223
Eastern red cedar
in Eastern revegetation programs, 216
in Midwestern revegetation programs, 218
Eastern revegetation programs
forbs in, 214
grasses in, 213–214
shrubs in, 215
trees in, 215–216
Eastern white pine
in Eastern revegetation programs, 216
in Midwestern revegetation programs, 219
Echinacea pallida, in Northern Great Plains revegetation programs, 221
Echinochloa crusgalli
in Eastern revegetation programs 213
in Midwestern revegetation programs, 217
Echinochloa crusgalli var. *frumentacea*, in Midwestern revegetation programs, 217
Ecosystem integrity, in applied reclamation, 1
Ecotones, landscape diversity and, 148
Ecotypes, plant species selection and, 115
Elaeagnus angustifolia
in Eastern revegetation programs, 216
in Midwestern revegetation programs, 218
in Northern Great Plains revegetation programs, 222
Elaeagnus umbellata
in Eastern revegetation programs, 215
in Midwestern revegetation programs, 218
Electrical conductivity
applied revegetation techniques and, 163–164
bentonite and, 194
chlorine and, 32
as soil salinity measure, 6–7
water-soluble boron and, 30
Electron transfer, iron in, 33
Elemental chemistry, plant and soil
introduction to, 15–17
macronutrients and, 17–24
Elevation
plant species selection and, 114–115
waste rock and, 171–174
Elymus angustus, in Northern Great Plains revegetation programs, 220
Elymus canadensis, in Northern Great Plains revegetation programs, 220
Elymus cinereus
in Northern Great Plains revegetation programs, 220
in Southwestern revegetation programs, 223
Elymus gigantus
in Northern Great Plains revegetation programs, 220
tailings and, 187
Elymus innovatus, in Northern Great Plains revegetation programs, 220

Elymus junceus
in Northern Great Plains revegetation programs, 220
in Southwestern revegetation programs, 223
Elymus salina, in Northern Great Plains revegetation programs, 220
Elymus triticoides, in Northern Great Plains revegetation programs, 220
Enlarged plow, plowing minesoil with, 70
Environmental Protection Agency (EPA), maximum tolerable lead levels approved by, 34
Enzymes
copper in, 32
iron in, 33
manganese in, 35
molybdenum in, 37
zinc in, 43
EPA; see Environmental Protection Agency
Ephedra spp., in Southwestern revegetation programs, 225
Eragrostis chloromelas, in Southwestern revegetation programs, 223
Eragrostis curvula
in Eastern revegetation programs, 213
in Midwestern revegetation programs, 217
in Southwestern revegetation programs, 223
Eragrostis intermedia, in Southwestern revegetation programs, 224
Eragrostis lehmanniana, in Southwestern revegetation programs, 223
Eragrostis superba, in Southwestern revegetation programs, 223
Eragrostis trichodes
in Midwestern revegetation programs, 217
in Southwestern revegetation programs, 223
Eragrostis trichophora, in Southwestern revegetation programs, 223
Eriogonum alatum, in Southwestern revegetation programs, 224
Eriogonum wrightii, in Southwestern revegetation programs, 225
Erodium cicutarium, in Southwestern revegetation programs, 224
Erosion
plant species selection and, 116
surface manipulation and, 61–63, 66, 70, 75–76
texture and, 10
Erosion control blankets, as surface mulch, 92–94
ESP; see Exchangeable sodium percentage
Eucalyptus camaldulensis
seeding and planting of, 111
in Southwestern revegetation programs, 226
Eucalyptus microtheca
seeding and planting of, 111
in Southwestern revegetation programs, 226
Europe, composting in, 95
European black alder
in Eastern revegetation programs, 215
in Midwestern revegetation programs, 218
European larch, in Eastern revegetation programs, 216
European white birch, in Eastern revegetation programs, 215
Eurotia lanata, in Southwestern revegetation programs, 225
Evergreen tamarix, in Southwestern revegetation programs, 226
Excelsior, in erosion control blankets, 92–93
Exchangeable sodium percentage (ESP), in estimation of sodium hazard, 7
Exposure
applied revegetation techniques and, 165
plant species selection and, 115
Extractants, soil, in determination of soil elemental levels, 15–16

F

Fagopyrum esculentum, in Eastern revegetation programs, 214
Fagopyrum sagittatum, in Eastern revegetation programs, 214
Fall seedings, characterization of, 127
False mesquite, in Southwestern revegetation programs, 225
Farming techniques, standard, in surface manipulation, 70–72
Federal guidelines
for maximum trace element loadings for agricultural lands, 97
on sewage sludge, 97
Federal Surface Mining Control and Reclamation Act, plant diversity requirements of, 151
Fertility, compost, 95

Fertilizers
fertilizer mechanics and, 84–85
introduction to, 79–80
mulches and, 85–94
nitrogen, 80–90
organic amendments and, 85–89, 94–100
phosphorus, 82–83
potassium, 83–84
Fertilizer spreader, in broadcast seeding, 122
Festuca arizonica, in Southwestern revegetation programs, 223
Festuca arundinacea
in Eastern revegetation programs, 213
in Midwestern revegetation programs, 217
in Northern Great Plains revegetation programs, 220
Festuca idahoensis
in Northern Great Plains revegetation programs, 220
waste rock dumps and, 175
Festuca ovina
in Northern Great Plains revegetation programs, 220
in Southwestern revegetation programs, 223
waste rock dumps and, 175
Festuca ovina var. *duriuscula*
in Northern Great Plains revegetation programs, 220
waste rock dumps and, 175
Festuca rubra
in Eastern revegetation programs, 213
in Midwestern revegetation programs, 217
in Northern Great Plains revegetation programs, 220
Festuca scabrella, in Northern Great Plains revegetation programs, 220
Festuca thruberi, in Northern Great Plains revegetation programs, 220
Field brome, in Eastern revegetation programs, 213
Field emergence, seed rates and, 117–118
Field soils, sulfur in, 23
Filaree, in Southwestern revegetation programs, 224
Final pit impoundments, with highwalls, 154–155
Fish and Game Department, state, impoundments and, 156
Fixed wing aircraft, seed dispersal by, 124
Flatpea, in Eastern revegetation programs, 214
Flood plains, drainage basin integrity and, 65
Flowering dogwood
in Eastern revegetation programs, 216
in Midwestern revegetation programs, 218
Fly ash, tailings and, 186–187
Food chains
cadmium in, 31
mercury in, 36
sewage sludge and, 97
Forage plants, selenium in, 41
Forbs
ash disposal ponds and, 203
bentonite mines and, 194
drainage basin integrity and, 65
in Eastern revegetation programs, 214
highway shoulders and, 200
iron in, 33
manganese in, 35
in Midwestern revegetation programs, 217–218
minesoils and, 166
in Northern Great Plains revegetation programs, 221
phosphorus fertilizers and, 83
in sand and gravel mines, 197
seed rates and, 118
slope reconstruction and, 61
in Southwestern revegetation programs, 224–225
tailings and, 187
waste rock dumps and, 174–175
Forestiera neomexicana, in Southwestern revegetation programs, 225
Forestiera pubescens, in Southwestern revegetation programs, 225
Forest Service, slope reconstruction recommendations of, 63
Fort Union formation, selenium in, 40
Fourwing saltbush
boron in, 29
minesoils and, 166–167
in Northern Great Plains revegetation programs, 222
in Southwestern revegetation programs, 225
Foxtail millet
in Eastern revegetation programs, 214
in Midwestern revegetation programs, 217
Fragrant sumac, in Eastern revegetation programs, 215
Franseria deltoidea, in Southwestern revegetation programs, 225

Fraxinus americana
in Eastern revegetation programs, 216
in Midwestern revegetation programs, 218
Fraxinus pennsylvanica
in Eastern revegetation programs, 216
in Midwestern revegetation programs, 218
in Northern Great Plains revegetation programs, 222
Fremont cottonwood, in Southwestern revegetation programs, 226
Fringed sagewort
minesoils and, 165–167
in Northern Great Plains revegetation programs, 222
in Southwestern revegetation programs, 225
waste rock dumps and, 174–175
Frost heaving, in planting failure, 136
Fungi
nitrogen fertilizers and, 81–82
soil organic matter and, 9
topsoiling and, 67
Fungicides, mercury in, 36
Fuzzy seeds, drill seeding problems and, 121

G

Galleta, in Southwestern revegetation programs, 223
Gamble oak, in Southwestern revegetation programs, 226
Gardner saltbush
bentonite mines and, 194
in Northern Great Plains revegetation programs, 222
Gels, against water deficit-induced transplant loss, 133–135
Generic spoil, topsoiling and, 69
Geomorphic stability, drainage basin integrity and, 65
Geomorphology, slope, 60–61
Germination, seed
applied revegetation techniques and, 163
broadcast seeding and, 122
drill seeding and, 121
erosion control blankets and, 94
irrigation and, 100
phosphorus and, 20
seeding techniques and, 118–119
seed rates and, 117–118
slope reconstruction and, 62
topsoiling and, 67
vegetation diversity and, 141–143, 147
Giant dropseed, in Southwestern revegetation programs, 224
Glycyrrhiza lepidota, in Northern Great Plains revegetation programs, 221
Golden currant, in Northern Great Plains revegetation programs, 222
Golden pea, in Northern Great Plains revegetation programs, 221
Goldenrod, in Northern Great Plains revegetation programs, 221
Golden Sunlight Mine, Montana, slope reconstruction at, 64
Gouging, in surface manipulation, 70, 73–78
Gradient, slope, 62–65
Granular soils, composition of, 10
Grass clippings, C:N ratio for, 89
Grasses
ash disposal ponds and, 203–204
bentonite mines and, 193–194
copper in, 32–33
drainage basin integrity and, 65
in Eastern revegetation programs, 213–214
highway shoulders and, 200
iron in, 33
manganese in, 35
in Midwestern revegetation programs, 217
minesoils and, 166
in Northern Great Plains revegetation programs, 219–221
phosphorus in, 20, 83
nitrogen in, 18
in sand and gravel mines, 197
seed rates and, 118
in Southwestern revegetation programs, 223–224
tailings and, 187
waste rock dumps and, 174–175
zinc in, 44
Grasslands
iron in, 34
manganese in, 35
nitrogen in, 81
topsoiling and, 69
Grass tetany, magnesium deficiency in, 23
Gravel pits, applied revegetation techniques and, 195–198
Gray birch, in Eastern revegetation programs, 215
Gray dogwood
in Eastern revegetation programs, 215

in Midwestern revegetation programs, 218
Grayia spinosa, in Northern Great Plains revegetation programs, 222
Greasewood, chlorine in, 31
Great Plains
arsenic and, 29
copper and, 33
fertilizers and, 79
grass tetany and, 23
iron and, 33–34
manganese and, 35
native hay mulches and, 92
organic matter and, 8
phosphorus fertilizers and, 84
recontouring and, 59
sulfur and, 23
wood residues and, 90
Green ash
in Eastern revegetation programs, 216
in Midwestern revegetation programs, 218
in Northern Great Plains revegetation programs, 222
Green manure, as organic amendment, 94–95
Green needlegrass
bentonite mines and, 193–194
minesoils and, 166
in Northern Great Plains revegetation programs, 115, 221
in Southwestern revegetation programs, 224
Green rabbitbush, in Northern Great Plains revegetation programs, 222
Green sagewort, in Northern Great Plains revegetation programs, 222
Green sprangletop, in Southwestern revegetation programs, 224
Grindelia spp., selenium in, 42
Ground cover, in plant succession, 2
Gully formation, slope reconstruction and, 61
Gumbo knots, inplace spoil ridges and, 157
Gutierrezia spp., selenium in, 42
Gypsum
in amelioration of sodicity, 99
bentonite mining and, 194–195
in western soils, 23

H

H^+, exchange capacity of, 12
Habitat
plant species selection and, 113
vegetation diversity and, 141
Hackberry
in Eastern revegetation programs, 216
in Midwestern revegetation programs, 218
in Northern Great Plains revegetation programs, 222
Hair grass, in Northern Great Plains revegetation programs, 219
Hairy goldenaster, in Northern Great Plains revegetation programs, 221
Hairy seeds, drill seeding problems and, 121
Hairy vetch
in Eastern revegetation programs, 214
in Midwestern revegetation programs, 218
Halfshrubs, minesoils and, 167
Halogeton glomeratus, sodium in, 43
Haplopappus spp., selenium in, 41
Hardening
bareroot stock and, 128–129
planting failure and, 136
Hardrock mines
abandoned wastes from, 206
acid-base balance and, 24
applied revegetation techniques and, 167–176
phosphorus and, 82
recontouring and, 59
slope reconstruction and, 63
soil pH and, 6
topsoiling and, 67
Hardsheep fescue
highway shoulders and, 200
tailings and, 187
waste rock dumps and, 174–176
Harrowing
surface manipulation and, 70, 72
tillage and, 112
Hawthorne
in Eastern revegetation programs, 215
in Midwestern revegetation programs, 218
Hay
community diversity and, 145, 147
as mulch, 91–92
phosphorus fertilizers and, 83
Hedysarum boreale, in Northern Great Plains revegetation programs, 221
Helianthus annuus
in Eastern revegetation programs, 214
in Midwestern revegetation programs, 217
in Northern Great Plains revegetation programs, 221

Helianthus maximiliani
in Eastern revegetation programs, 214
in Midwestern revegetation programs, 217
in Northern Great Plains revegetation programs, 221
Helicopters, seed dispersal by, 124
Herbicides
arsenic in, 27
highway shoulders and, 200–201
scalping and, 128, 137
Heterotheca villosa, in Northern Great Plains revegetation programs, 221
Hickory
in Eastern revegetation programs, 215
in Midwestern revegetation programs, 218
High-bush cranberry, in Eastern revegetation programs, 215
Highwalls
bluff extensions and, 151
impoundments and, 154–155
Highway drainage, sodded, 135
Highway shoulders
applied revegetation techniques and, 198–201
upper soil horizon storage and, 18
Hilaria jamesii
in Northern Great Plains revegetation programs, 220
in Southwestern revegetation programs, 223
Hillslopes
drainage basin integrity and, 64–66
metal mine wastes and, 169
reconstruction of, 60–62
Holding pond, tailings sluiced to, 179, 182
Hookers evening primrose, in Southwestern revegetation programs, 225
Horizontal juniper
in Northern Great Plains revegetation programs, 222
in sand and gravel mines, 197–198
Horse manure, C:N ratio for, 89
Hot water extraction
of boron, 17, 30
of molybdenum, 37
of selenium, 40, 42
of zinc, 16
Humus, as soil organic matter, 8
Hydrochloric acid extraction
of arsenic, 28
of lead, 35
Hydrogen ions
calcium and, 21, 26–27
lime and, 97
plant establishment and, 5–6
tailings and, 180–181
Hydrologic properties, improving, amendments and, 76
Hydromulcher, in mulch application, 85–86
Hydromulching
site preparation and, 64, 86–89
waste rock dumps and, 172
Hydroseeding
characterization of, 122–124
surface manipulation and, 74
waste rock dumps and, 172

I

Idaho
grass tetany in, 23
phosphate mining in, 188–190
waste rock slopes in, 175
Illinois bundleflower, in Eastern revegetation programs, 214
Illinois Coal Basin, impoundments in, 154
Impoundments, landscape diversity and, 152–156
Incubation techniques, in determining nitrogen mineralization potential, 19
Indiangrass
in Eastern revegetation programs, 214
in Midwestern revegetation programs, 217
in Southwestern revegetation programs, 224
Indian ricegrass
minesoils and, 166
in sand and gravel mines, 197
in Southwestern revegetation programs, 224
waste rock dumps and, 175
Indigobush, in Eastern revegetation programs, 215
Infiltration
compost and, 95
drainage basin integrity and, 66
organic matter and, 8
slope reconstruction and, 61–62, 64
surface manipulation and, 75–76
topsoiling and, 66, 69
waste rock slopes and, 171
Inplace spoil ridges, landscape diversity and, 156–157
Insecticides, arsenic in, 28–29

Insects
organic matter and, 8
topsoiling and, 67
Intermediate wheatgrass
in Northern Great Plains revegetation programs, 219
in Southwestern revegetation programs, 223
Interseeding, community diversity and, 145, 147–148
Invading species, vegetation diversity and, 142–143, 146
Invertebrates
nitrogen fertilizers and, 82
topsoiling and, 68
Iowa, tall grass prairie in, 1
Iron
calcium and, 34
coal ash and, 201
deficiency of, 25, 34
functions of, 33
ionic forms of, 16
nickel and, 39
toxicity of, 25, 34
Iron pyrite
lime and, 97
oxidation of, 13
Irrigation, in site preparation, 100
Islands, sod, community diversity and, 145, 147
Italian ryegrass
in Eastern revegetation programs, 213
in Midwestern revegetation programs, 217

J

Jack pine
in Eastern revegetation programs, 216
in Midwestern revegetation programs, 218
Japanese fleeceflower, in Eastern revegetation programs, 214
Japanese honeysuckle, in Eastern revegetation programs, 215
Japanese larch, in Eastern revegetation programs, 216
Japanese lespedeza, in Eastern revegetation programs, 215
Japanese millet
in Eastern revegetation programs, 213
in Midwestern revegetation programs, 217
Johnsongrass, in Southwestern revegetation programs, 224
Jojoba, in Southwestern revegetation programs, 226
"J" roots, planting failure and, 137
Juglans nigra
in Eastern revegetation programs, 216
in Midwestern revegetation programs, 218
Juniperus communis, in Northern Great Plains revegetation programs, 222
Juniperus horizontalis
in Northern Great Plains revegetation programs, 222
in sand and gravel mines, 197–198
Juniperus osteosperma, in Southwestern revegetation programs, 226
Juniperus spp., as containerized stock, 130
Juniperus virginiana
in Eastern revegetation programs, 216
in Midwestern revegetation programs, 218
Jute, in erosion control blankets, 92–94

K

K^+, exchange capacity of, 12
Kentucky bluegrass
in Eastern revegetation programs, 214
in Midwestern revegetation programs, 217
minesoils and, 165
seeding and planting of, 110
sodding and, 135
in Southwestern revegetation programs, 224
Kentucky "31" tall fescue
in Eastern revegetation programs, 213
in Midwestern revegetation programs, 217
King Ranch bluestem, in Southwestern revegetation programs, 223
Kjeldahl procedure, total nitrogen determined via, 19
Kleingrass, in Southwestern revegetation programs, 224
Kobe lespedeza
in Eastern revegetation programs, 214
in Midwestern revegetation programs, 217
Kochia prostrata, in Northern Great Plains revegetation programs, 221
Koeleria cristata
in Northern Great Plains revegetation programs, 220
in Southwestern revegetation programs, 224
Koeleria pyramidata
minesoils and, 165

in Northern Great Plains revegetation programs, 220
in Southwestern revegetation programs, 224
Korean lespedeza
in Eastern revegetation programs, 214
in Midwestern revegetation programs, 217

L

Ladino clover
in Eastern revegetation programs, 214
in Midwestern revegetation programs, 218
Lake Superior, tailings in, 178
Landscape diversity, enhancement of, 148–158
Larix decidua, in Eastern revegetation programs, 216
Larix leptolepis, in Eastern revegetation programs, 216
Larrea tridentata, in Southwestern revegetation programs, 225
Later-successional species, rehabilitation and, 110
Lathyrus sylvestris, in Eastern revegetation programs, 214
Latuca sativa, aluminum in, 26
Leached soils
nitrate ions and, 18
sulfur deficiency in, 23
topsoiling and, 69
Lead
calcium and, 21
coal ash and, 201
deficiency of, 25
federal recommendations for maximum loading of in agricultural lands, 97
sewage sludge and, 96
tailings and, 180
topsoiling and, 68
toxicity of, 25, 34–35
Lead plant
in Midwestern revegetation programs, 217
in Northern Great Plains revegetation programs, 221
Leaf lettuce, aluminum in, 26
Legislation, of reclamation, 58
Legumes
copper in, 33
green manure and, 95
molybdenum in, 37
nitrogen deficiency in, 18
phosphorus fertilizers and, 83
seeding of, 124–125
Lehman's lovegrass, in Southwestern revegetation programs, 223
Leptochloa dubia, in Southwestern revegetation programs, 224
Lespedeza bicolor
in Eastern revegetation programs, 215
in Midwestern revegetation programs, 218
Lespedeza cuneata
in Eastern revegetation programs, 214
in Midwestern revegetation programs, 217
Lespedeza daurica, in Eastern revegetation programs, 214
Lespedeza japonica, in Eastern revegetation programs, 215
Lespedeza stipulacea
in Eastern revegetation programs, 214
in Midwestern revegetation programs, 217
Lespedeza striata
in Eastern revegetation programs, 214
in Midwestern revegetation programs, 217
Lespedeza thunbergii, in Eastern revegetation programs, 215
Letterman needlegrass, in Northern Great Plains revegetation programs, 221
Lewis' flax
in Northern Great Plains revegetation programs, 221
in Southwestern revegetation programs, 224
Liatris punctata, in Northern Great Plains revegetation programs, 221
Liatris pycnostachya, in Eastern revegetation programs, 214
Life cycle, plant, stages of, 4
Light, germination and, 118
Ligustrum amurense, in Eastern revegetation programs, 215
Lime
Bray extraction and, 20
as calcium amendment, 97–99
tailings and, 179–181, 185–187
Limestone; see Lime
Liming
boron and, 30
calcium and, 22
magnesium and, 23
sulfur and, 24
Linum lewisii
in Northern Great Plains revegetation programs, 221

in Southwestern revegetation programs, 224
Linum perenne, minesoils and, 167
Liquidambar styraciflua
in Eastern revegetation programs, 216
in Midwestern revegetation programs, 218
Liriodendron tulipifera
in Eastern revegetation programs, 216
in Midwestern revegetation programs, 218
Lithologic control, of erosion, 61
Little bluestem
in Eastern revegetation programs, 214
in Midwestern revegetation programs, 217
minesoils and, 166
in Southwestern revegetation programs, 224
Little leaf Palo verde, in Southwestern revegetation programs, 225
Livestock waste, as organic amendment, 94–95
Loam soils
porosity of, 12
seed rates and, 118
Loblolly pine
in Eastern revegetation programs, 216
in Midwestern revegetation programs, 219
Locust, in Eastern revegetation programs, 215
Lodgepole pine, in Northern Great Plains revegetation programs, 222
Lodorm green needlegrass, in northern Great Plains revegetation programs, 115
Lolium multiflorum
in Eastern revegetation programs, 213
in Midwestern revegetation programs, 217
Lolium perenne
in Eastern revegetation programs, 213
in Midwestern revegetation programs, 217
in Northern Great Plains revegetation programs, 220
Longleaf pine, in Eastern revegetation programs, 216
Lonicera japonica, in Eastern revegetation programs, 215
Lonicera maackii, in Eastern revegetation programs, 215
Lonicera morrowi, in Eastern revegetation programs, 215
Lonicera tatarica, in Eastern revegetation programs, 215
Lotus corniculatus
in Eastern revegetation programs, 214
in Midwestern revegetation programs, 217
in Northern Great Plains revegetation programs, 221
tailings and, 187
Lotus tenuis, in Eastern revegetation programs, 214
Lupinus sericeus, in Northern Great Plains revegetation programs, 221
Lupinus spp., waste rock dumps and, 176

M

Machaeranthera spp., selenium in, 42
Maclura pomifera
in Eastern revegetation programs, 216
in Midwestern revegetation programs, 218
Macrohabitats, vegetation diversity and, 142
Macronutrients; see specific nutrients
Macropores, soil, function of, 12
Magnesium
calcium and, 21–22
coal ash and, 201
deficiency of, 22–23
ionic forms of, 16
sodium and, 7, 43
toxicity of, 22–23
Magnesium chloride
bentonite mining and, 194
sodicity and, 99
Malus spp., in Eastern revegetation programs, 216
Mammoth wildrye, tailings and, 187
Management practices, community diversity and, 145, 148
Manganese
calcium and, 21
deficiency of, 25
ionic forms of, 16
tailings and, 180
topsoiling and, 68
toxicity of, 25, 35–36
Mantle, soil
bluff extensions and, 152
tailings and, 184
Manure
C:N ratio for, 89
as organic amendment, 8, 94
tailings and, 183
Manure spreader, in manure application, 85–86, 94
Marcasite, tailings and, 180
Marginal leaf burning, chlorine in, 32

Mat saltbush, in Southwestern revegetation programs, 225
Maximilian sunflower
  in Eastern revegetation programs, 214
  in Midwestern revegetation programs, 217
  in Northern Great Plains revegetation programs, 221
MCS; see Metal Concentration in the Sludge
Meadow foxtail, in Northern Great Plains revegetation programs, 219
Medicago lupulina, in Northern Great Plains revegetation programs, 221
Medicago sativa
  in Eastern revegetation programs, 214
  in Midwestern revegetation programs, 217
  nitrogen deficiency in, 18
  in Northern Great Plains revegetation programs, 221
  in Southwestern revegetation programs, 225
Mediterranium grass, in Southwestern revegetation programs, 224
Melilotus alba
  copper in, 33
  in Eastern revegetation programs, 214
  in Midwestern revegetation programs, 217
  in Northern Great Plains revegetation programs, 221
  in Southwestern revegetation programs, 225
Melilotus officinalis
  in Eastern revegetation programs, 214
  in Midwestern revegetation programs, 217
  in Northern Great Plains revegetation programs, 221
  in sand and gravel mines, 197
  in Southwestern revegetation programs, 225
  waste rock dumps and, 176
Melilotis spp., minesoils and, 167
Membrane liner, coal ash disposal pond with, 202
Menodora scabra, in Southwestern revegetation programs, 225
Mercury, toxicity of, 36–37
Mesa dropseed, in Southwestern revegetation programs, 224
Metal Concentration in the Sludge (MCS), for sewage sludge, 97
Metal concentrators, pollution caused by, arsenic in, 29
Metal mine wastes, applied revegetation techniques and, 167–188
Metal reduction works, mercury and, 36
Metalloenzymes, plant, copper in, 32
Metals; see also specific metals
  sewage sludge and, 96–97
  soil pH and, 5
Mexican hat, in Southwestern revegetation programs, 225
Mexican Palo verde, in Southwestern revegetation programs, 225
Mexico, soil pH of mine sites in, 6
Mg^{2+}, exchange capacity of, 12
Microhabitats, vegetation diversity and, 141–142
Micronutrients; see specific nutrients
Microorganisms
  organic matter and, 9
  topsoiling and, 67
Micropores, soil, function of, 12
Mill wastes, phosphorus fertilizers and, 84
Mineral soils, barren, slope reconstruction and, 63
Minesoils
  amendments and, 98–99
  arsenic and, 29
  boron and, 30
  broadcast-seeded plant community on, 123
  cadmium and, 31
  chiselling and, 77
  copper and, 32–33
  dozer basins and, 75, 77
  drainage basin integrity and, 65
  drill-seeded plant community on, 121
  gouging and, 73–74, 77–78
  green manure and, 95
  impoundments and, 152–153
  infiltration rates and, 61
  iron pyrite and, 24
  lead and, 35
  mercury and, 37
  nickel and, 39
  phosphorus fertilizers and, 83
  recontouring and, 59–60
  rock ledges and, 157–158
  selenium and, 42
  soil pH and, 6
  slope reconstruction and, 61–64
  standard farming techniques and, 70–72
  topsoiling and, 67, 69
Minespoils, abandoned, 206
Mine wastes, applied revegetation techniques and, 167–198
Minnesota, tailings in, 178

Moisture, soil
plant species selection and, 116
seed rates and, 118
slope reconstruction and, 62
Moldboard plow, in surface manipulation, 70
Molybdenosis, induced copper deficiency in, 37–38
Molybdenum
applied revegetation techniques and, 164
copper and, 33, 37–38
deficiency of, 25, 37
ionic forms of, 16
toxicity of, 25, 37–38
Montana
acid-base balance in soils of, 24
bentonite mining in, 194
cadmium in, 31
community diversity and, 145, 148
copper in, 32
impoundments in, 152, 155
lead in, 35
manganese in, 35
molybdenum in, 37
nickel in, 39
nitrogen fertilizers and, 81
phosphate mining in, 188
seed mixes and, 112
slope reconstruction in, 61, 64
sodic soil amendments in, 99
waste rock slopes in, 175
zinc in, 43
Mormon tea, in Southwestern revegetation programs, 225
Morrow honeysuckle, in Eastern revegetation programs, 215
Morus rubra, in Eastern revegetation programs, 216
Mountain brome
phosphate mines and, 191
in Southwestern revegetation programs, 223
waste rock dumps and, 174–175
Mountain mahogany, in Southwestern revegetation programs, 225
Mountain muhly, in Southwestern revegetation programs, 224
Mountains
phosphorus fertilizers and, 84
waste rock and, 171–174
Muhlenbergia montana, in Southwestern revegetation programs, 224
Muhlenbergia porteri, in Southwestern revegetation programs, 224
Muhlenbergia wrightii, in Southwestern revegetation programs, 224
Mulberry, in Eastern revegetation programs, 216
Mulches
bentonite mines and, 193
community diversity and, 145, 147
site preparation and, 8, 85–94
Multiple land use goals, plant species selection and, 116
Multiple seeding dates, community diversity and, 145, 148
Mycorrhizal associations
rehabilitation and, 110
topsoiling and, 67–69

N

Na^+, exchange capacity of, 12
Narrowleaf trefoil, in Eastern revegetation programs, 214
Native hay, as mulch, 91–92, 145, 147
Native plants, definition of, 110
Needle-and-thread grass
ash disposal ponds and, 203–204
bentonite mines and, 193
iron in, 33
minesoils and, 165–166
in Northern Great Plains revegetation programs, 221
in sand and gravel mines, 197
in Southwestern revegetation programs, 224
Netting
erosion control blankets and, 94
in securing mulches, 86, 91
Neutrality, soil pH and, 6
New Guinea, tailings in, 178
New Mexico locust, in Southwestern revegetation programs, 226
New Mexico olive, in Southwestern revegetation programs, 225
Niches, vegetation diversity and, 141, 144
Nickel
federal recommendations for maximum loading of in agricultural lands, 97
iron and, 39
magnesium and, 22
sewage sludge and, 96
tailings and, 180
toxicity of, 22, 38–39
Niobrara silt loams, selenium in, 39

Nitrate
ammonium ions and, 18
fertilizers and, 80–81
molybdenum and, 37
Nitrite, soil concentrations of, 19
Nitrogen
amendments and, 88–90
compost and, 95–96
cycle, 18, 81–82
deficiency of, 18
fertilizers and, 80–82, 84–85, 89–90
fixation of by legumes, 124–125
green manure and, 95
as macronutrient, 16, 18–20
manure and, 95
minesoils and, 167
molybdenum and, 37
nutrient cycle rehabilitation and, 9–10
tailings and, 181
zinc and, 43
North Dakota, topsoiling in, 67
Northern Catalpa, in Eastern revegetation programs, 216
Northern red oak
in Eastern revegetation programs, 216
in Midwestern revegetation programs, 219
Northern sweetvetch, in Northern Great Plains revegetation programs, 221
Norway spruce, in Eastern revegetation programs, 216
Nucleic acid synthesis, zinc in, 43
Nurse crops
ash disposal ponds and, 204
tailings and, 187
temporary stabilization and, 125–126
waste rock dumps and, 176
Nut pine, in Southwestern revegetation programs, 226
Nutrient pools, plant, topsoiling and, 67
Nutrient reserves, soil, organic matter-induced increase of, 8

O

Oat stems, in mulch, 90
Oat straw, C:N ratio for, 89
Oenothera hookerii, in Southwestern revegetation programs, 225
Oldman wormwood, in Northern Great Plains revegetation programs, 222
Olsen extraction, of phosphorus, 20
Onobrychis viceaefolia
in Eastern revegetation programs, 214
in Midwestern revegetation programs, 218
minesoils and, 167
in Northern Great Plains revegetation programs, 221
Open-pit mines
applied revegetation techniques and, 167–169
impoundments and, 154–155
Opportunists
plant succession and, 2
rehabilitation and, 109
Opuntia polyacantha, minesoils and, 165
Orchard grass
in Eastern revegetation programs, 213
in Midwestern revegetation programs, 217
phosphate mines and, 191
waste rock dumps and, 174
Orchard soils, arsenic in, 27, 29
Ore bodies, mercury and, 36
Oregon, grass tetany in, 23
Ores, metallic, arsenic in, 29
Organic amendments
site preparation and, 85–89, 94–100
tailings and, 183–184
Organic matter
amendments and, 76
green manure and, 95
manure and, 95
nitrogen and, 18
phosphorus and, 20
soil and, 8–10
sulfur and, 23
topsoiling and, 67–68
Oryzopsis hymenoides
minesoils and, 166
in Northern Great Plains revegetation programs, 220
in Southwestern revegetation programs, 224
Osage orange
in Eastern revegetation programs, 216
in Midwestern revegetation programs, 218
Overburden, topsoiling and, 68–69
Oxidation-reduction reactions, copper in, 32
Oxidized rock, in slope reconstruction, 64

P

Painted milkvetch, in Northern Great Plains revegetation programs, 221
Panicum antidotale, in Southwestern revegetation programs, 224

Panicum clandestinum
in Eastern revegetation programs, 213
in Midwestern revegetation programs, 217
Panicum coloratum, in Southwestern revegetation programs, 224
Panicum miliaceum, in Eastern revegetation programs, 213
Panicum obtusum, in Southwestern revegetation programs, 224
Panicum virgatum
in Eastern revegetation programs, 213
in Midwestern revegetation programs, 217
in Northern Great Plains revegetation programs, 220
in Southwestern revegetation programs, 224
Paper, as mulch, 89
Paper birch, in Eastern revegetation programs, 215
Parkinsonia aculeata, in Southwestern revegetation programs, 225
Partridge pea, in Eastern revegetation programs, 214
Paspalum dilatatum, in Eastern revegetation programs, 213
Paspalum notatum, in Eastern revegetation programs, 213
Paulownia tomentosa, in Eastern revegetation programs, 216
Peabody Coal Company, impoundments and, 152
Pea plants, zinc in, 43
Pearl millet
in Eastern revegetation programs, 213
in Midwestern revegetation programs, 217
Pecan
in Eastern revegetation programs, 215
in Midwestern revegetation programs, 218
Peking cotoneaster, in Northern Great Plains revegetation programs, 222
Pennisetum americanum
in Eastern revegetation programs, 213
in Midwestern revegetation programs, 217
Pennsylvania
acid slopes in, 186
sewage sludge utilization in, 96
Penstemon spp.
in Eastern revegetation programs, 214
in Northern Great Plains revegetation programs, 221
selenium in, 42
in Southwestern revegetation programs, 225
Penstemon strictus, in Southwestern revegetation programs, 225
Percolation, high permeability and, 12
Permeability, soil
rates of, 12
texture influence on, 10
Perennial plants, in temporary stabilization, 125–126
Perennial ryegrass
in Eastern revegetation programs, 213
in Midwestern revegetation programs, 217
Pesticides, arsenic in, 27
Petalostemon purpureum, in Southwestern revegetation programs, 225
pH, soil
aluminum and, 26–27
amendments and, 76, 79
hydrogen ions 5
lead and, 35
lime and, 98
molybdenum and, 37
nickel and, 38
plant species selection and, 113
sewage sludge and, 96
topsoiling and, 67–69
Phalaris arundinacea
in Eastern revegetation programs, 213
in Midwestern revegetation programs, 217
in Northern Great Plains revegetation programs, 220
Phleum alpinum, in Northern Great Plains revegetation programs, 220
Phleum pratense
drill seeding of, 120
in Eastern revegetation programs, 213
in Midwestern revegetation programs, 217
in Northern Great Plains revegetation programs, 220
Phosphate extraction, of arsenic, 29
Phosphate mining, applied revegetation techniques and, 188–191
Phosphogypsum
bentonite mining and, 194
sodicity and, 99
Phosphoric acid, in fertilizers, 83
Phosphorus
arsenic and, 28
deficiency of, 20, 82–83
fertilizers and, 82–83
as macronutrient, 16, 18, 20
nutrient cycle rehabilitation and, 9
sewage sludge and, 96

soil levels of, 83
soil pH and, 5
tailings and, 181, 186
toxicity of, 20
Phosphorus pentoxide, in fertilizers, 84–85
Photosystem II, manganese in, 35
Phragmites communis, in Northern Great Plains revegetation programs, 220
Picea abies, in Eastern revegetation programs, 216
Picia glauca, in Eastern revegetation programs, 216
Pin oak
in Eastern revegetation programs, 216
in Midwestern revegetation programs, 219
Pine dropseed, in Southwestern revegetation programs, 223
Pinus banksiana
in Eastern revegetation programs, 216
in Midwestern revegetation programs, 218
Pinus cembroides, in Southwestern revegetation programs, 226
Pinus contorta, in Northern Great Plains revegetation programs, 222
Pinus echinata
in Eastern revegetation programs, 216
in Midwestern revegetation programs, 218
Pinus edulis, in Southwestern revegetation programs, 226
Pinus elliotti, in Eastern revegetation programs, 216
Pinus monophylla, in Southwestern revegetation programs, 226
Pinus nigra
in Eastern revegetation programs, 216
in Midwestern revegetation programs, 218
Pinus palustria, in Eastern revegetation programs, 216
Pinus ponderosa
boron in, 29
in Northern Great Plains revegetation programs, 223
in Southwestern revegetation programs, 226
Pinus resinosa
in Eastern revegetation programs, 216
in Midwestern revegetation programs, 218
Pinus rigida
in Eastern revegetation programs, 216
in Midwestern revegetation programs, 219
Pinus strobus
in Eastern revegetation programs, 216
in Midwestern revegetation programs, 219
Pinus sylvestris
in Eastern revegetation programs, 216
in Midwestern revegetation programs, 219
Pinus taeda
in Eastern revegetation programs, 216
in Midwestern revegetation programs, 219
Pinus virginiana
in Eastern revegetation programs, 216
in Midwestern revegetation programs, 219
Pinyon pine, in Southwestern revegetation programs, 226
Pioneer species
plant succession and, 2
rehabilitation and, 109
temporary stabilization and, 125
Pipeline corridors, upper soil horizon storage and, 18
Pitch pine
in Eastern revegetation programs, 216
in Midwestern revegetation programs, 219
Pitting, with large depressions, in slope reconstruction, 64
Placer mine waste, abandoned, 205
Plains bristlegrass, in Southwestern revegetation programs, 224
Plains lovegrass, in Southwestern revegetation programs, 223
Plant communities
broadcast-seeded, 122–123
drill-seeded, 120–121
Plant fragments, topsoiling and, 67
Planting
planting failure and, 136–137
of plant parts, 133
special planting techniques and, 133–135
of whole plants, 128–132
Plant Materials Centers
phosphate mines and, 191
plant species selection and, 115
seed rates and, 117
Plant parts, transplantation of, 128, 133
Plant species selection, for seed mixes, 113–117
Plant successional communities
characterization of, 2
drainage basin integrity and, 66
slope reconstruction and, 62
Plant successional species, rehabilitation and, 109–111
Platanus occidentalis
in Eastern revegetation programs, 216

in Midwestern revegetation programs, 219
Plowing
of lime into acid minesoil, 98–99
phosphorus fertilizers and, 83
surface manipulation and, 70
tillage and, 112
PLS; see Pure live seed
Poa alpina, in Northern Great Plains revegetation programs, 220
Poa ampla, in Northern Great Plains revegetation programs, 220
Poa bulbosa, in Northern Great Plains revegetation programs, 220
Poa canbyi, in Northern Great Plains revegetation programs, 220
Poa compressa
in Eastern revegetation programs, 214
in Midwestern revegetation programs, 217
in Northern Great Plains revegetation programs, 220
tailings and, 187
Poa pratensis
in Eastern revegetation programs, 214
in Midwestern revegetation programs, 217
in Northern Great Plains revegetation programs, 220
seeding and planting of, 110
in Southwestern revegetation programs, 224
Poa sandbergii
in Northern Great Plains revegetation programs, 220
in Southwestern revegetation programs, 224
Poa secunda
in Northern Great Plains revegetation programs, 220
in Southwestern revegetation programs, 224
Pollution, nickel and, 39; see also Air pollution
Polygonum cuspidatum, in Eastern revegetation programs, 214
Polymers, against water deficit-induced transplant loss, 133–135
Ponderosa pine
boron in, 29
community diversity and, 145
in Northern Great Plains revegetation programs, 223
in Southwestern revegetation programs, 226
Populus deltoides
in Eastern revegetation programs, 216
in Midwestern revegetation programs, 219
in Northern Great Plains revegetation programs, 223
Populus fremontii, in Southwestern revegetation programs, 226
Populus tremuloides, in Northern Great Plains revegetation programs, 223
Porosity, of soil, 8, 10–12
Postdisturbance plant community, applied revegetation techniques and, 162
Postmine land use, applied revegetation techniques and, 161
Potassium
calcium and, 21
deficiency of, 21, 83–84
fertilizers and, 83–84
as macronutrient, 16, 18, 21
sewage sludge and, 96
soil levels of, 84
tailings and, 181
toxicity of, 21
Potassium chloride
fertilizers and, 84
nitrates and, 19
Potassium nitrate, in fertilizers, 84
Potassium oxide, in fertilizers, 84–85
Potassium sulfate, in fertilizers, 84
Potentilla fruticosa, in Northern Great Plains revegetation programs, 222
Powder River Coal Basin, Wyoming
applied revegetation techniques and, 164
bluff extensions and, 151
rock ledges and, 158
Prairie adelia, in Southwestern revegetation programs, 225
Prairie blazing star, in Eastern revegetation programs, 214
Prairie coneflower
in Eastern revegetation programs, 214
in Northern Great Plains revegetation programs, 221
Prairie junegrass
minesoils and, 165
in Southwestern revegetation programs, 224
waste rock dumps and, 175
Prairie rose, in Northern Great Plains revegetation programs, 222

Prairie sandreed grass
  minesoils and, 166
  in sand and gravel mines, 197
Prairie soils
  acid-base balance and, 24
  infiltration rates and, 61
  restoration of, 1
  zinc in, 43
Precipitation
  applied revegetation techniques and, 163
  compost and, 95
  drainage basin integrity and, 65
  infiltration and, 76
  macropore action during, 12
  planting failure and, 136
  plant species selection and, 114–115
  polymers and, 135
  season of seeding and, 127
  slope reconstruction and, 61–62
  surface modification and, 77
Predisturbance plant community
  applied revegetation techniques and, 162
  plant species selection and, 113–115
Previous use, definition of, 110
Prickly pear cactus, minesoils and, 165
Prickly rose, in Northern Great Plains revegetation programs, 222
Propagules, plant, topsoiling and, 67–69
Proso millet, in Eastern revegetation programs, 213
Prosopis alba, in Southwestern revegetation programs, 225
Prosopis chilensis, in Southwestern revegetation programs, 225
Prosopis juliflora, in Southwestern revegetation programs, 225
Prosopis velutina, in Southwestern revegetation programs, 225
Prostrate lespedeza, in Eastern revegetation programs, 214
Prostrate summercypress, in Northern Great Plains revegetation programs, 221
Proteins
  sulfur in, 23
  zinc in, 43
Protozoans
  nitrogen fertilizers and, 81–82
  topsoiling and, 67
Prunus americana
  in Eastern revegetation programs, 215
  in Northern Great Plains revegetation programs, 222
Prunus besseyi, in Eastern revegetation programs, 215
Prunus pumila, in Northern Great Plains revegetation programs, 222
Prunus serotina
  in Eastern revegetation programs, 216
  in Midwestern revegetation programs, 219
Prunus virginiana
  in Eastern revegetation programs, 215
  in Northern Great Plains revegetation programs, 222
Pseudotsuga menziesii
  in Eastern revegetation programs, 216
  in Northern Great Plains revegetation programs, 223
Pubescent wheatgrass
  bentonite mines and, 194
  in Northern Great Plains revegetation programs, 219
  in Southwestern revegetation programs, 223
  tailings and, 187
Puccinellia distans, in Northern Great Plains revegetation programs, 220
Pure live seed (PLS), percent, seed rates and, 117–118
Purple osier willow, in Eastern revegetation programs, 215
Purple prairie clover
  ash disposal ponds and, 203–204
  bentonite mines and, 194
  in Eastern revegetation programs, 214
  highway shoulders and, 200
  minesoils and, 166–167
  in Northern Great Plains revegetation programs, 221
  in sand and gravel mines, 197
  in Southwestern revegetation programs, 225
Purple threeawn, in Southwestern revegetation programs, 223
Purshia tridentata
  in Northern Great Plains revegetation programs, 222
  in Southwestern revegetation programs, 226
Pyritic materials
  acid-generating potential of, 24
  tailings and, 179–180, 186

Q

Quailbush, in Southwestern revegetation programs, 225
Quaking aspen, in Northern Great Plains revegetation programs, 223
Quercus acutissima
  in Eastern revegetation programs, 216
  in Midwestern revegetation programs, 219
Quercus alba
  in Eastern revegetation programs, 216
  in Midwestern revegetation programs, 219
Quercus gambelii, in Southwestern revegetation programs, 226
Quercus imbricaria
  in Eastern revegetation programs, 216
  in Midwestern revegetation programs, 219
Quercus macrocarpa
  in Eastern revegetation programs, 216
  in Midwestern revegetation programs, 219
  in Northern Great Plains revegetation programs, 222
Quercus palustris
  in Eastern revegetation programs, 216
  in Midwestern revegetation programs, 219
Quercus prinus, in Midwestern revegetation programs, 219
Quercus rubra
  in Eastern revegetation programs, 216
  in Midwestern revegetation programs, 219

R

Raindrop impact, slope reconstruction and, 61
Range grasses
  copper in, 33
  nitrogen in, 18
Rangelands
  chlorine in, 32
  compost and, 95
  copper in, 32
  recontouring and, 60
  salinity of, 7
  selenium in, 41–42
Rangeland sod, manipulation of, 135
Raspberry, plant parts and, 133
Ratibida columnaris, in Southwestern revegetation programs, 225
Ratibida columnifera
  in Eastern revegetation programs, 214
  in Northern Great Plains revegetation programs, 221
Reclamation, definition of, 1
Recommended Maximum Metal Loading Rate (RMMLR), for sewage sludge, 97
Recontouring
  applied revegetation techniques and, 162–163
  site preparation and, 58–66
Recruitment, plant, vegetation diversity and, 143
Red clover
  in Eastern revegetation programs, 214
  in Midwestern revegetation programs, 218
  in Northern Great Plains revegetation programs, 221
  waste rock dumps and, 174
Red fescue
  in Eastern revegetation programs, 213
  in Midwestern revegetation programs, 217
Red gum, in Southwestern revegetation programs, 226
Red maple
  in Eastern revegetation programs, 215
  in Midwestern revegetation programs, 218
Red-osier dogwood
  in Eastern revegetation programs, 215
  in Midwestern revegetation programs, 218
  in Northern Great Plains revegetation programs, 222
Red pine
  in Eastern revegetation programs, 216
  in Midwestern revegetation programs, 219
Redtop
  in Eastern revegetation programs, 213
  in Midwestern revegetation programs, 217
  in Northern Great Plains revegetation programs, 219
  tailings and, 180
Reed canarygrass
  in Eastern revegetation programs, 213
  in Midwestern revegetation programs, 217
Regreen, as nurse crop, 126
Regulatory restrictions
  applied revegetation techniques and, 163–164
  on plant species selection, 113–114, 116–117
Rehabilitation, definition of, 1
Restoration, definition of, 1
Revegetation
  definition of, 1
  implementation of, 109–110

Rhus aromatica, in Eastern revegetation programs, 215
Rhus copallina
in Eastern revegetation programs, 215
in Midwestern revegetation programs, 218
Rhus trilobata
iron in, 33
in Northern Great Plains revegetation programs, 222
in Southwestern revegetation programs, 226
Ribes aureum, in Northern Great Plains revegetation programs, 222
Ribes cereum
in Northern Great Plains revegetation programs, 222
in Southwestern revegetation programs, 226
Rilling, slope reconstruction and, 61
Riparian habitats
drainage basin integrity and, 65
impoundments and, 156
Ripping
compaction and, 177
surface manipulation and, 70, 72–73
River birch
in Eastern revegetation programs, 215
in Midwestern revegetation programs, 218
RMMLR; see Recommended Maximum Metal Loading Rate
Robinia fertilis, in Eastern revegetation programs, 215
Robinia hispida
in Eastern revegetation programs, 215
in Midwestern revegetation programs, 218
Robinia neomexicana, in Southwestern revegetation programs, 226
Robinia pseudoacacia
in Eastern revegetation programs, 216
in Midwestern revegetation programs, 219
Rock ledges, landscape diversity and, 157–158
Rocky mountain penstemon, in Southwestern revegetation programs, 225
Root pads, in establishing shrubs or trees, 133
Roots, plant
green manure and, 95
organic matter and, 8
slope reconstruction and, 62
plant succession and, 2
topsoiling and, 69
Rootzone
amendments and, 79
applied revegetation techniques and, 162–164
ash disposal ponds and, 203
lime and, 97
plant species selection and, 114
plant succession and, 2
recontouring and, 59
selenium and, 42
sewage sludge and, 96
slope reconstruction and, 62
tailings and, 183–186
Rosa acicularis, in Northern Great Plains revegetation programs, 222
Rosa arkansana, in Northern Great Plains revegetation programs, 222
Rosana western wheatgrass, in northern Great Plains revegetation programs, 115
Rosa rugosa, in Eastern revegetation programs, 215
Rosa spp., plant parts and, 133
Rosa woodsii
in Northern Great Plains revegetation programs, 222
in Southwestern revegetation programs, 226
Rough menodora, in Southwestern revegetation programs, 225
Royal paulownia, in Eastern revegetation programs, 216
Rubber rabbitbush
minesoils and, 167
in Northern Great Plains revegetation programs, 222
in Southwestern revegetation programs, 225
Rubus spp., plant parts and, 133
Rugosa rose, in Eastern revegetation programs, 215
Runoff
drainage basin integrity and, 65–66
slope reconstruction and, 62
Russian olive
in Eastern revegetation programs, 216
in Midwestern revegetation programs, 218
in Northern Great Plains revegetation programs, 222
Russian soils, arsenic in, 28
Russian thistle, as pioneer species, 125–126
Russian wildrye, in Southwestern revegetation programs, 223

S

Sagebrush
community diversity and, 145
minesoils and, 167
waste rock dumps and, 176
Sainfoin
in Eastern revegetation programs, 214
in Midwestern revegetation programs, 218
minesoils and, 167
in Northern Great Plains revegetation programs, 221
Salix bebbiana, in Northern Great Plains revegetation programs, 223
Salix exigua, in Northern Great Plains revegetation programs, 223
Salix nigra, in Eastern revegetation programs, 216
Salix purpurea, in Eastern revegetation programs, 215
Salix spp.
in Eastern revegetation programs, 216
manganese in, 35
in Northern Great Plains revegetation programs, 222
Salsola kali, as pioneer species, 125–126
Saltcedar, in Southwestern revegetation programs, 226
Saltgrass
chlorine in, 31
in Southwestern revegetation programs, 223
Salts, topsoiling and, 69
Salvia azurea, in Eastern revegetation programs, 214
Sambucus canadensis, in Eastern revegetation programs, 215
Sandbar willow, in Northern Great Plains revegetation programs, 223
Sandburg bluegrass
in Southwestern revegetation programs, 224
waste rock dumps and, 175
Sand cherry, in Northern Great Plains revegetation programs, 222
Sand dropseed
in Northern Great Plains revegetation programs, 221
in Southwestern revegetation programs, 224
Sand lovegrass
in Midwestern revegetation programs, 217
in Southwestern revegetation programs, 223
Sand pits, applied revegetation techniques and, 195–198
Sandy clay, composition of, 10
Sandy-loam soils, plant species selection and, 114
Sandy soils
boron in, 30
diameter range of, 10
sulfur in, 23
Sanguisorba minor
in Northern Great Plains revegetation programs, 221
in Southwestern revegetation programs, 225
SAR; see Sodium absorption ratio
Sarcobatus vermiculatus
chlorine in, 31
in Northern Great Plains revegetation programs, 222
in Southwestern revegetation programs, 226
Saturated paste, soil pH values determined in, 6
Sawdust
C:N ratio for, 89
as mulch, 89–90
Sawtooth oak
in Eastern revegetation programs, 216
in Midwestern revegetation programs, 219
Scalping, in removal of direct root competition, 128, 137
Scarlet globemallow, in Northern Great Plains revegetation programs, 221
Schismus barbatus, in Southwestern revegetation programs, 224
Schizachyrium scoparium
in Eastern revegetation programs, 214
in Midwestern revegetation programs, 217
minesoils and, 166
in Northern Great Plains revegetation programs, 220
in Southwestern revegetation programs, 224
Scotch pine
as bareroot stock, 129
in Eastern revegetation programs, 216
in Midwestern revegetation programs, 219
Season
applied revegetation techniques and, 163, 165–166

plant species selection and, 114
Sediment yields, drainage basin integrity and, 65–66
Seedbed preparation, in surface manipulation, 72–78, 111–112
Seeding techniques, characterization of, 118–124; see also specific techniques
Seedling establishment, topsoiling and, 67–68
Seed mixes, designing, 112–118
Seed rates
community diversity and, 145
unit area of soil and, 117–118
waste rock dumps and, 176
Seeds
plant succession and, 2
topsoiling and, 67
Seed sterile hybrids, as nurse crops, 126
Selenate, identification of, 40
Selenides, in overburden material, 42
Selenites, in overburden material, 42
Selenium
applied revegetation techniques and, 164
coal ash and, 201
deficiency of, 25, 40–41
oxidation and, 17
topsoiling and, 68
toxicity of, 25, 40–42
volatility of, 39
Seral species
plant succession and, 2
rehabilitation and, 109–110
Sericea lespedeza
in Eastern revegetation programs, 214
in Midwestern revegetation programs, 217
Serpentine soils
calcium in, 22
magnesium in, 22
nickel in, 38–39
Serviceberry, in Northern Great Plains revegetation programs, 222
Setaria italica
in Eastern revegetation programs, 214
in Midwestern revegetation programs, 217
Setaria macrostachya, in Southwestern revegetation programs, 224
Sewage sludge
C:N ratio for, 89
as organic amendment, 96–97
tailings and, 186
Shadscale
in Northern Great Plains revegetation programs, 222
in Southwestern revegetation programs, 225
Sheep fescue
in Southwestern revegetation programs, 223
waste rock dumps and, 175–176
Shepherdia argentea
in Eastern revegetation programs, 215
in Northern Great Plains revegetation programs, 222
Shingle oak
in Eastern revegetation programs, 216
in Midwestern revegetation programs, 219
Shining sumac
in Eastern revegetation programs, 215
in Midwestern revegetation programs, 218
Shortleaf pine
in Eastern revegetation programs, 216
in Midwestern revegetation programs, 218
Shrubby buckwheat, in Southwestern revegetation programs, 225
Shrubby cinquefoil, in Northern Great Plains revegetation programs, 222
Shrub lespedeza
in Eastern revegetation programs, 215
in Midwestern revegetation programs, 218
Shrub planters, mechanized, 131–132
Shrubs
bentonite mines and, 194
drainage basin integrity and, 65
in Eastern revegetation programs, 215
iron in, 33
manganese in, 35
in Midwestern revegetation programs, 218
minesoils and, 166
in Northern Great Plains revegetation programs, 221–222
in sand and gravel mines, 197
seed rates and, 118
slope reconstruction and, 61
in Southwestern revegetation programs, 225
waste rock dumps and, 174–175
Siberian peashrub
in Eastern revegetation programs, 215
in Midwestern revegetation programs, 218
in Northern Great Plains revegetation programs, 222
Siberian wheatgrass
in Northern Great Plains revegetation programs, 219
in Southwestern revegetation programs, 223

Sideoats grama
in Eastern revegetation programs 213
minesoils and, 166
in Southwestern revegetation programs, 223
Silicon, coal ash and, 201
Silky dogwood
in Eastern revegetation programs, 215
in Midwestern revegetation programs, 218
Silky lupine, in Northern Great Plains revegetation programs, 221
Silt, diameter range of, 10
Silty-clay soils, plant species selection and, 114
Silver buffaloberry
in Eastern revegetation programs, 215
in Northern Great Plains revegetation programs, 222
Silver maple
in Eastern revegetation programs, 215
in Midwestern revegetation programs, 218
Silver sagebrush
minesoils and, 165
in Northern Great Plains revegetation programs, 222
Simmondsia chinensis, in Southwestern revegetation programs, 226
Single leaf pinyon, in Southwestern revegetation programs, 226
Sinuosity, drainage basin integrity and, 65
Sitanion hystrix, in Northern Great Plains revegetation programs, 221
Site preparation
introduction to, 57–58
recontouring and, 58–66
Skunkbush sumac
community diversity and, 145
iron in, 33
minesoils and, 165
in Northern Great Plains revegetation programs, 222
in Southwestern revegetation programs, 226
Slag, tailings and, 187
Slash pine, in Eastern revegetation programs, 216
Slender wheatgrass
ash disposal ponds and, 203–204
bentonite mines and, 194
minesoils and, 166
in Northern Great Plains revegetation programs, 219
in Southwestern revegetation programs, 223
tailings and, 187
waste rock dumps and, 175–176
Slope reconstruction, recontouring and, 60–64
Slopes
applied revegetation techniques and, 163, 165
bluff extensions and, 151–152
highway shoulders and, 198
hydroseeding and, 123
metal mine wastes and, 169
seed rates and, 118
Small burnet
in Northern Great Plains revegetation programs, 221
in Southwestern revegetation programs, 225
SMCRA; see Surface Mining Control and Reclamation Act
Smelters
arsenic and, 29
phosphorus and, 84
Smooth brome
in Eastern revegetation programs, 213
manganese in, 35
in Midwestern revegetation programs, 217
minesoils and, 164
phosphate mines and, 191
Snowberry, plant parts and, 133
Snowbrush ceanothus
in Northern Great Plains revegetation programs, 222
waste rock dumps and, 175–176
Snow melt, drainage basin integrity and, 65
Soapweed, in Northern Great Plains revegetation programs, 222
Sodding
community diversity and, 145, 147
disturbed land revegetation and, 135
Sodic soils
amendments and, 99–100
bentonite and, 192
calcium and, 21
sodium and, 7
topsoiling and, 69
Sodium
ash disposal ponds and, 202
alkalinity and, 7–8
bentonite and, 191
calcium and, 43

coal ash and, 201
magnesium and, 43
topsoiling and, 69
Sodium absorption ratio (SAR), in estimation of sodium hazard, 7–8, 22–23, 68–69, 99, 194
Soil
B concentration in, 13
buffering capacity of, 13
bulk density and porosity of, 10–12
cation exchange capacity in, 12–13
hydrogen ion concentration in, 5–7
mineralogy of, 13
organic matter in, 8–10
oxidation-reduction potential of, 13
sodium in, 7–8
structure of, 10
texture and structure of, 10
Soil Conservation Service
applied revegetation techniques and, 162
drill seeding recommendations of, 120
Soil decomposer organisms, in soil organic matter, 9
Soil paste
boron in, 30
calcium in, 22
chlorine in, 32
magnesium in, 22
sodium in, 43
Soil solution
cation exchange capacity and, 13
organic matter and, 9
Soil-plant-animal system, relationships between physical and chemical environment of, 3
Solidago missouriensis, in Northern Great Plains revegetation programs, 221
Sorghastrum nutans
in Eastern revegetation programs, 214
in Midwestern revegetation programs, 217
in Southwestern revegetation programs, 224
Sorghum bicolor, in Eastern revegetation programs, 214
Sorghum halapense, in Southwestern revegetation programs, 224
Sorghum sudanense
in Eastern revegetation programs, 214
in Midwestern revegetation programs, 217
Sorghum vulgare
in Eastern revegetation programs, 214
in Midwestern revegetation programs, 217
South Africa, tailings in, 178
South Dakota, bentonite mining in, 191–194
Southern arrowood, in Eastern revegetation programs, 215
Sphaeralcea coccinea, in Northern Great Plains revegetation programs, 221
Sphaeralcea grossulariafolia, in Southwestern revegetation programs, 225
Spike muhly, in Southwestern revegetation programs, 224
Spiny hopsage, in Northern Great Plains revegetation programs, 222
Sporobolus airoides
in Northern Great Plains revegetation programs, 221
in Southwestern revegetation programs, 224
Sporobolus cryptandrus
in Northern Great Plains revegetation programs, 221
in Southwestern revegetation programs, 224
Sporobolus flexuosus, in Southwestern revegetation programs, 224
Sporobolus giganteus, in Southwestern revegetation programs, 224
Sprigging, in establishing shrubs or trees, 133
Spring seedings, characterization of, 127
Sprinkler irrigation, on mined lands, 100
Stability, soil
plant species selection and, 116
slope reconstruction and, 60
Stanleya spp., selenium in, 41
Stansbury cliffrose, in Southwestern revegetation programs, 225
State certifying agencies, seed rates and, 117
State regulatory agencies, slope reconstruction and, 63
Stipa columbiana
in Northern Great Plains revegetation programs, 221
waste rock dumps and, 175
Stipa comata
iron in, 33
in Northern Great Plains revegetation programs, 221
in Southwestern revegetation programs, 224
Stipa lettermanii, in Northern Great Plains revegetation programs, 221

Stipa viridula
in Northern Great Plains revegetation programs, 115, 221
in Southwestern revegetation programs, 224
Straw
bentonite mines and, 193
C:N ratio for, 89
in erosion control blankets, 92–94
as mulch, 8, 85, 87, 90–91
Streambank wheatgrass
bentonite mines and, 194
highway shoulders and, 200
in Northern Great Plains revegetation programs, 219
in sand and gravel mines, 197
in Southwestern revegetation programs, 223
Stream channels, drainage basin integrity and, 64–66
Stripmines
applied revegetation techniques and, 161–165
impoundments and, 153, 155
recontouring and, 59
rock ledges and, 157
slope reconstruction and, 63
Stubble mulch, in site preparation, 92
Stubble mulch crops, temporary stabilization and, 126
Subterranean clover, in Eastern revegetation programs, 214
Sudangrass
in Eastern revegetation programs, 214
in Midwestern revegetation programs, 217
Sugar maple, in Midwestern revegetation programs, 218
Sugar translocation and utilization, boron in, 29
Sulfur
deficiency of, 23–24
impoundments and, 155
ionic forms of, 16
lime and, 98
nutrient cycle rehabilitation and, 9
oxidation and, 17
toxicity of, 23
Sulfur dioxide, fumigation with, 23
Sulfuric acid extraction
of arsenic, 28
iron pyrite and, 24
Superphosphate, in fertilizers, 83
Surface manipulation, site preparation and, 69–76
Surface Mining Control and Reclamation Act (SMCRA)
applied revegetation techniques and, 161–163
enactment of, 4
plant species selection and, 114
recontouring and, 59
terms used in, 110–111
topsoiling and, 66
Surface soil, topsoiling and, 66, 69
Sweet birch, in Eastern revegetation programs, 215
Sweetgum
in Eastern revegetation programs, 216
in Midwestern revegetation programs, 218
Switchgrass
in Eastern revegetation programs, 213
in Midwestern revegetation programs, 217
in Southwestern revegetation programs, 224
Symphoricarpos occidentalis
in Northern Great Plains revegetation programs, 222
in Southwestern revegetation programs, 226
Symphoricarpos spp., plant parts and, 133

T

Tackifiers
hydroseeding and, 123
securing mulches and, 64, 86–91
Tailings, applied revegetation techniques and, 178–188
Tall grass prairie, restoration of, 1
Tall oatgrass, in Eastern revegetation programs, 213
Tall wheatgrass
in Northern Great Plains revegetation programs, 219
in Southwestern revegetation programs, 223
Tamarix aphylla, in Southwestern revegetation programs, 226
Tamarix pentandra, in Southwestern revegetation programs, 226
Tatarian honeysuckle, in Eastern revegetation programs, 215
Taxodium distichum
in Eastern revegetation programs, 216

in Midwestern revegetation programs, 219
Temperature
amendments and, 76
applied revegetation techniques and, 163
plant species selection and, 114, 116
Temporary stabilizing species, seeding, 125–126
Textural triangle, soil, 11
Texture
bentonite and, 192
plant species selection and, 113–114
Thermopsis rhombifolia, in Northern Great Plains revegetation programs, 221
Thickspike gayfeather, in Northern Great Plains revegetation programs, 221
Thickspike wheatgrass
in Northern Great Plains revegetation programs, 115, 219
in sand and gravel mines, 197
in Southwestern revegetation programs, 223
tailings and, 187
Thixotropic materials, tailings and, 182
Threadleaf sedge, minesoils and, 165
Thunberg lespedeza, in Eastern revegetation programs, 215
Ticklegrass, in Northern Great Plains revegetation programs, 219
Tilia americana, in Eastern revegetation programs, 216
Tillage
seedbed preparation and, 112
surface manipulation and, 72
Timothy
drill seeding of, 120
in Eastern revegetation programs, 213
in Midwestern revegetation programs, 217
phosphate mines and, 191
waste rock dumps and, 174
Tissues, plant, in plant succession, 2
Topsoil
applied revegetation techniques and, 162–163, 165
bentonite and, 192–193
community diversity and, 146–147
highway shoulders and, 198
landscape diversity and, 150–151
metal mine wastes and, 169
seedbed preparation and, 111–112
seed rates and, 118
sewage sludge and, 96
slope reconstruction and, 63–64
Topsoiled acid spoils, grass production and, 69
Topsoiling, site preparation and, 66–69
Trace elements; see specific elements
Transplantation
into disturbed soil, 128–133
planting failure and, 136
Tree-of-heaven
in Midwestern revegetation programs, 218
in Southwestern revegetation programs, 226
Tree planters, mechanized, 131–132
Trees
drainage basin integrity and, 65
in Eastern revegetation programs, 215–217
in Midwestern revegetation programs, 218–219
in Northern Great Plains revegetation programs, 222
seed rates and, 118
slope reconstruction and, 61
in Southwestern revegetation programs, 226
Tree spade, mechanized, in transplantation of wildings, 130–131
Triangle leaf bursage, in Southwestern revegetation programs, 225
Trichachne californica, in Southwestern revegetation programs, 224
Trifolium hybridum
in Eastern revegetation programs, 214
in Midwestern revegetation programs, 218
in Northern Great Plains revegetation programs, 221
Trifolium incarnatum, in Eastern revegetation programs, 214
Trifolium pratense
in Eastern revegetation programs, 214
in Midwestern revegetation programs, 218
in Northern Great Plains revegetation programs, 221
Trifolium repens
in Eastern revegetation programs, 214
in Midwestern revegetation programs, 218
in Northern Great Plains revegetation programs, 221
Trifolium subterraneum, in Eastern revegetation programs, 214
Triple superphosphate, in fertilizers, 83
Triticum aestivum, in mulch, 90

Tufted hairgrass
in Eastern revegetation programs, 213
in Midwestern revegetation programs, 217
in Northern Great Plains revegetation programs, 115
tailings and, 180

U

Underground mining, applied revegetation techniques and, 161
Urban centers, cadmium in, 31
USDA; see U.S. Department of Agriculture
U.S. Department of Agriculture (USDA), Handbook No. 60 of, 6
U.S. Geological Survey
arsenic and, 28
boron and, 30
mercury and, 36
zinc and, 44
Utah, phosphate mining in, 188
Utah juniper, in Southwestern revegetation programs, 226
Utah serviceberry, in Southwestern revegetation programs, 225

V

Vaccinium spp., in Eastern revegetation programs, 215
Valley floor, drainage basin integrity and, 65–66
Variable depth topsoiling, landscape diversity and, 150–151
Vegetation performance, pH and, 6
Velvet mesquite, in Southwestern revegetation programs, 225
Vernalization, seeding and, 127
Viburnum dentatum, in Eastern revegetation programs, 215
Viburnum opulus, in Eastern revegetation programs, 215
Vicia americana, in Northern Great Plains revegetation programs, 221
Vicia sativa, in Eastern revegetation programs, 214
Vicia villosa
in Eastern revegetation programs, 214
in Midwestern revegetation programs, 218
Vigna unguiculata, in Eastern revegetation programs, 214
Vine mesquite, in Southwestern revegetation programs, 224
Virginia pine
in Eastern revegetation programs, 216
in Midwestern revegetation programs, 219

W

Walkley Black method, in soil organic carbon content determination, 9
Washington, waste rock slopes in, 175
Waste dumps, slope reconstruction and, 63
Waste rock, applied revegetation techniques and, 170–176
Water, impoundments and, 152–156
Water deficit, in transplant loss, 133–136
Water holding capacity
access roads and, 176
compost and, 95
organic matter and, 8
Watersheds, drainage basin integrity and, 64–66
Watersplitting, in photosystem II, manganese in, 35
Water tables
copper in, 33
gravel pit with, 196
interception of, 195
Wax currant
in Northern Great Plains revegetation programs, 222
in Southwestern revegetation programs, 226
Weathering, natural, drainage basin integrity and, 64
Weathering chambers, lime and, 98
Weeds
highway shoulders and, 200–201
nitrogen fertilizers and, 81
seed rates and, 118
slope reconstruction and, 63
Weeping lovegrass
in Eastern revegetation programs, 213
in Midwestern revegetation programs, 217
in Southwestern revegetation programs, 223
Western Energy Coal Mine
community diversity and, 145
inplace spoil ridges and, 157
Western sandcherry, in Eastern revegetation programs, 215
Western snowberry
in Northern Great Plains revegetation programs, 222

in Southwestern revegetation programs, 226
Western wheatgrass
ash disposal ponds and, 203
bentonite mines and, 194
drill seeding of, 120
highway shoulders and, 200
minesoils and, 165–166
in Northern Great Plains revegetation programs, 115, 219
in Southwestern revegetation programs, 223
Western yarrow
minesoils and, 165–167
in Northern Great Plains revegetation programs, 221
in Southwestern revegetation programs, 224
waste rock dumps and, 174–176
Wetlands
impoundments and, 155–156
restoration of, 1
Wheat stems, in mulch, 90
Wheat straw, C:N ratio for, 89
White ash
in Eastern revegetation programs, 216
in Midwestern revegetation programs, 218
White clover
in Eastern revegetation programs, 214
in Midwestern revegetation programs, 218
in Northern Great Plains revegetation programs, 221
White muscle disease, selenium deficiency in, 41
White oak
in Eastern revegetation programs, 216
in Midwestern revegetation programs, 219
White spruce, in Eastern revegetation programs, 216
White sweetclover
copper in, 33
in Eastern revegetation programs, 214
in Midwestern revegetation programs, 218
molybdenum in, 37
in Northern Great Plains revegetation programs, 221
in Southwestern revegetation programs, 225
Whitethorn, in Southwestern revegetation programs, 225
Whole plants, transplantation of, 128–132
Wild buckwheat, in Southwestern revegetation programs, 224
Wildings, transplantation of, 128, 130–131, 136
Wild licorice, in Northern Great Plains revegetation programs, 221
Wild plum, in Northern Great Plains revegetation programs, 222
Wild roses, plant parts and, 133
Willow
in Eastern revegetation programs, 216
manganese in, 35
in Northern Great Plains revegetation programs, 223
plant parts and, 133
Wilman lovegrass, in Southwestern revegetation programs, 223
Winterfat
minesoils and, 166–167
in Northern Great Plains revegetation programs, 222
in Southwestern revegetation programs, 225
Wood chips, in soil organic matter, 8, 183, 193
Wood fibers, in erosion control blankets, 92–93
Wood residues, as mulch, 89–90
Worms, topsoiling and, 67
Wyoming
acid-base analyses required by, 24
bentonite mining in, 191, 194
broadcast seeding in, 124
drill seeding in, 124
phosphate mining in, 188
rock ledges and, 158
selenium in, 39
sodic soil amendments in, 99
wood residues and, 90

X

Xylorrhiza spp., selenium in, 41

Y

Yellow bluestem, in Southwestern revegetation programs, 223
Yellow poplar
in Eastern revegetation programs, 216
in Midwestern revegetation programs, 218
Yellowstone National Park
seeding of native plants in, 174
topsoiling and, 68–69

Yellow sweetclover
ash disposal ponds and, 203–204
bentonite mines and, 194
in Eastern revegetation programs, 214
in Midwestern revegetation programs, 218
in Northern Great Plains revegetation programs, 221
in Southwestern revegetation programs, 225
tailings and, 187
waste rock dumps and, 174, 176
Yucca glauca, in Northern Great Plains revegetation programs, 222

Z

Zinc
ammonium acetate extraction of, 16
deficiency of, 43
federal recommendations for maximum loading of in agricultural lands, 97
ionic forms of, 16
sewage sludge and, 96
tailings and, 180
topsoiling and, 68
toxicity of, 43–44